Use your online resources to get the best out of your prep.

**Get started at
kaptest.com/moreonline**

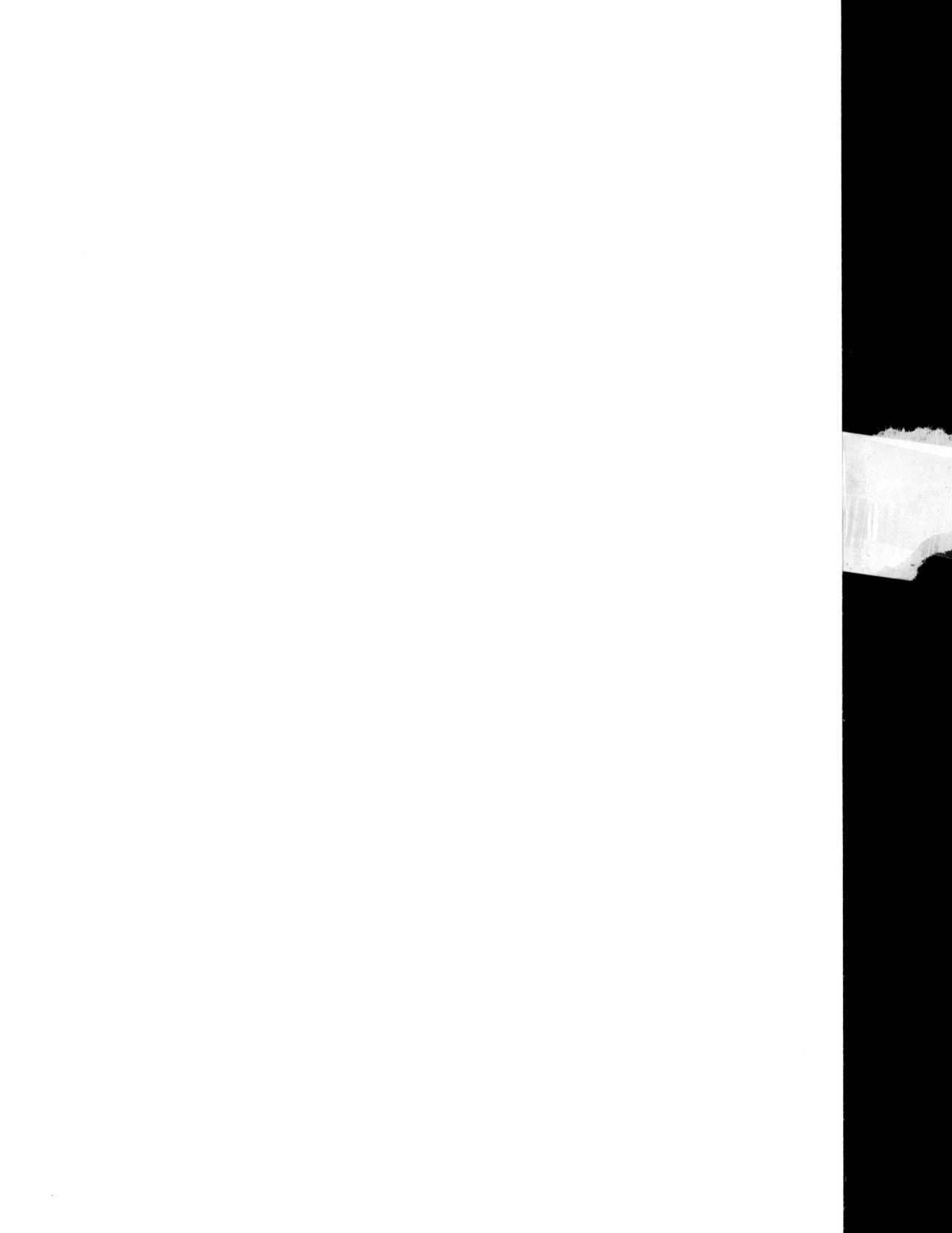

MCAT® | Critical Analysis and Reasoning Skills Review

2019–2020

Edited by Alexander Stone Macnow, MD

PUBLISHING

New York

MCAT® is a registered trademark of the Association of American Medical Colleges, which neither sponsors nor endorses this product.

This publication is designed to provide accurate and authoritative information in regard to the subject matter covered. It is sold with the understanding that the publisher is not engaged in rendering medical, legal, accounting, or other professional services. If legal advice or other expert assistance is required, the services of a competent professional should be sought.

© 2018 by Kaplan, Inc.

Published by Kaplan Publishing, a division of Kaplan, Inc.
750 Third Avenue
New York, NY 10017

All rights reserved under International and Pan-American Copyright Conventions. By payment of the required fees, you have been granted the non-exclusive, non-transferable right to access and read the text of this eBook on screen. No part of this text may be reproduced, transmitted, downloaded, decompiled, reverse engineered, or stored in or introduced into any information storage and retrieval system, in any form or by any means, whether electronic or mechanical, now known or hereinafter invented, without the express written permission of the publisher.

ISBN: 978-1-5062-3540-0

10 9 8 7 6 5 4 3 2 1

Kaplan Publishing print books are available at special quantity discounts to use for sales promotions, employee premiums, or educational purposes. For more information or to purchase books, please call the Simon & Schuster special sales department at 866-506-1949.

Preface

And now it starts: your long, yet fruitful journey toward wearing a white coat. Proudly wearing that white coat, though, is hopefully only part of your motivation. You are reading this book because you want to be a healer.

If you're serious about going to medical school, then you are likely already familiar with the importance of the MCAT in medical school admissions. While the holistic review process puts additional weight on your experiences, extracurricular activities, and personal attributes, the fact remains: along with your GPA, your MCAT score remains one of the two most important components of your application portfolio—at least early in the admissions process. Each additional point you score on the MCAT pushes you in front of thousands of other students and makes you an even more attractive applicant. But the MCAT is not simply an obstacle to overcome; it is an opportunity to show schools that you will be a strong student and a future leader in medicine.

We at Kaplan take our jobs very seriously and aim to help students see success not only on the MCAT, but as future physicians. We work with our learning science experts to ensure that we're using the most up-to-date teaching techniques in our resources. Multiple members of our team hold advanced degrees in medicine or associated biomedical sciences, and are committed to the highest level of medical education. Kaplan has been working with the MCAT for over 50 years and our commitment to premed students is unflagging; in fact, Stanley Kaplan created this company when he had difficulty being accepted to medical school due to unfair quota systems that existed at the time.

We stand now at the beginning of a new era in medical education. As citizens of this 21st-century world of healthcare, we are charged with creating a patient-oriented, culturally competent, cost-conscious, universally available, technically advanced, and research-focused healthcare system, run by compassionate providers. Suffice it to say, this is no easy task. Problem-based learning, integrated curricula, and classes in interpersonal skills are some of the responses to this demand for an excellent workforce—a workforce of which you'll soon be a part.

We're thrilled that you've chosen us to help you on this journey. Please reach out to us to share your challenges, concerns, and successes. Together, we will shape the future of medicine in the United States and abroad; we look forward to helping you become the doctor you deserve to be.

Good luck!

Alexander Stone Macnow, MD
Editor-in-Chief
Department of Pathology and Laboratory Medicine
Hospital of the University of Pennsylvania

BA, Musicology—Boston University, 2008
MD—Perelman School of Medicine at the University of Pennsylvania, 2013

Table of Contents

Preface..iii
The *Kaplan MCAT Review* Team vi
About the MCAT...vii
How This Book Was Created..xviii
Using This Book..xix

Chapter 1: About CARS — 1
   1.1 The CARS Section 4
   1.2 Passages 4
   1.3 Question Categories 5

Chapter 2: Analyzing Rhetoric — 9
   2.1 What Is Rhetoric? 12
   2.2 Key Components of Rhetoric...12
   2.3 Rhetorical Analysis 17

Chapter 3: Keywords — 25
   3.1 Reading Strategically with Keywords.....................................28
   3.2 Relation Keywords...30
   3.3 Author Keywords...34
   3.4 Logic Keywords 37

Chapter 4: Outlining the Passage — 57
   4.1 The Kaplan Method for CARS Passages................................60
   4.2 Reverse-Engineering the Author's Outline...........................64
   4.3 Practicing the Strategy 68

Chapter 5: Dissecting Arguments — 95
   5.1 Domains of Discourse...98
   5.2 Concepts: The Basic Elements of Logic 100
   5.3 Claims: The Bearers of Truth Value..102
   5.4 Arguments: Conclusions and Evidence 103

Additional resources available at www.kaptest.com/mcatbookresources

Chapter 6: Formal Logic — 111
- 6.1 The Logic of Conditionals — 114
- 6.2 Applications of Conditionals — 117
- 6.3 Analogical Reasoning — 120

Chapter 7: Understanding Passages — 125
- 7.1 Varieties of Passages — 128
- 7.2 Support in Passages — 133
- 7.3 Anticipating Questions — 136

Chapter 8: Question and Answer Strategy — 161
- 8.1 Kaplan Method for CARS Questions — 164
- 8.2 Wrong Answer Pathologies — 168
- 8.3 Signs of a Healthy Answer — 170

Chapter 9: Question Types I: *Foundations of Comprehension* Questions — 193
- 9.1 Main Idea Questions — 196
- 9.2 Detail Questions — 198
- 9.3 Function Questions — 203
- 9.4 Definition-in-Context Questions — 205

Chapter 10: Question Types II: *Reasoning Within the Text* Questions — 227
- 10.1 Inference Questions — 230
- 10.2 Strengthen–Weaken (Within the Passage) Questions — 236
- 10.3 Other *Reasoning Within the Text* Questions — 242

Chapter 11: Question Types III: *Reasoning Beyond the Text* Questions — 263
- 11.1 Apply Questions — 266
- 11.2 Strengthen–Weaken (Beyond the Passage) Questions — 273
- 11.3 Other *Reasoning Beyond the Text* Questions — 277

Chapter 12: Effective Review of CARS — 301
- 12.1 Learning from Your Mistakes — 304
- 12.2 Thinking Like the Testmaker: Post-Phrasing — 306
- 12.3 Improving Your Timing — 308
- 12.4 Building Endurance — 310
- 12.5 Enhancing Your Vocabulary — 310

The *Kaplan MCAT Review* Team

Alexander Stone Macnow, MD
Editor-in-Chief

Áine Lorié, PhD
Editor

Kristen L. Russell, ME
Editor

Derek Rusnak, MA
Editor

Pamela Willingham, MSW
Editor

Mikhail Alexeeff
Kaplan MCAT Faculty

Melinda Contreras, MS
Kaplan MCAT Faculty

Laura L. Ambler
Kaplan MCAT Faculty

Samantha Fallon
Kaplan MCAT Faculty

Krista L. Buckley, MD
Kaplan MCAT Faculty

Jason R. Pfleiger
Kaplan MCAT Faculty

Faculty Reviewers and Editors: Elmar R. Aliyev; James Burns; Jonathan Cornfield; Alisha Maureen Crowley; Christopher Durland; Nikolai Dorofeev, MD; Benjamin Downer, MS; Colin Doyle; M. Dominic Eggert; Marilyn Engle; Eleni M. Eren; Raef Ali Fadel; Elizabeth Flagge; Adam Grey; Tyra Hall-Pogar, PhD; Scott Huff; Samer T. Ismail; Elizabeth A. Kudlaty; Kelly Kyker-Snowman, MS; Ningfei Li; John P. Mahon; Matthew A. Meier; Nainika Nanda; Caroline Nkemdilim Opene; Kaitlyn E. Prenger; Uneeb Qureshi; Bela G. Starkman, PhD; Michael Paul Tomani, MS; Nicholas M. White; Allison Ann Wilkes, MS; Kerranna Williamson, MBA; and Tony Yu

Thanks to Kim Bowers; Eric Chiu; Tim Eich; Tyler Fara; Owen Farcy; Dan Frey; Robin Garmise; Rita Garthaffner; Joanna Graham; Allison Harm; Beth Hoffberg; Aaron Lemon-Strauss; Keith Lubeley; Diane McGarvey; Petros Minasi; John Polstein; Deeangelee Pooran-Kublall, MD, MPH; Rochelle Rothstein, MD; Larry Rudman; Sylvia Tidwell Scheuring; Carly Schnur; Karin Tucker; Lee Weiss; and the countless others who made this project possible.

About the MCAT

ANATOMY OF THE MCAT

Here is a general overview of the structure of Test Day:

Section	Number of Questions	Time Allotted
Test-Day Certification		4 minutes
Tutorial (optional)		10 minutes
Chemical and Physical Foundations of Biological Systems	59	95 minutes
Break (optional)		10 minutes
Critical Analysis and Reasoning Skills (CARS)	53	90 minutes
Lunch Break (optional)		30 minutes
Biological and Biochemical Foundations of Living Systems	59	95 minutes
Break (optional)		10 minutes
Psychological, Social, and Biological Foundations of Behavior	59	95 minutes
Void Question		3 minutes
Satisfaction Survey (optional)		5 minutes

The structure of the four sections of the MCAT is shown below.

Chemical and Physical Foundations of Biological Systems	
Time	95 minutes
Format	• 59 questions • 10 passages • 44 questions are passage-based, and 15 are discrete (stand-alone) questions. • Score between 118 and 132
What It Tests	• Biochemistry: 25% • Biology: 5% • General Chemistry: 30% • Organic Chemistry: 15% • Physics: 25%
Critical Analysis and Reasoning Skills (CARS)	
Time	90 minutes
Format	• 53 questions • 9 passages • All questions are passage-based. There are no discrete (stand-alone) questions. • Score between 118 and 132
What It Tests	Disciplines: • Humanities: 50% • Social Sciences: 50% Skills: • *Foundations of Comprehension*: 30% • *Reasoning Within the Text*: 30% • *Reasoning Beyond the Text*: 40%

Biological and Biochemical Foundations of Living Systems	
Time	95 minutes
Format	• 59 questions • 10 passages • 44 questions are passage-based, and 15 are discrete (stand-alone) questions. • Score between 118 and 132
What It Tests	• Biochemistry: 25% • Biology: 65% • General Chemistry: 5% • Organic Chemistry: 5%
Psychological, Social, and Biological Foundations of Behavior	
Time	95 minutes
Format	• 59 questions • 10 passages • 44 questions are passage-based, and 15 are discrete (stand-alone) questions. • Score between 118 and 132
What It Tests	• Biology: 5% • Psychology: 65% • Sociology: 30%
Total	
Testing Time	375 minutes (6 hours, 15 minutes)
Total Seat Time	447 minutes (7 hours, 27 minutes)
Questions	230
Score	472 to 528

SCIENTIFIC INQUIRY AND REASONING SKILLS (SIRS)

The AAMC has defined four *Scientific Inquiry and Reasoning Skills* (SIRS) that will be tested in the three science sections of the MCAT:

1. *Knowledge of Scientific Concepts and Principles* (35% of questions)
2. *Scientific Reasoning and Problem-Solving* (45% of questions)
3. *Reasoning About the Design and Execution of Research* (10% of questions)
4. *Data-Based and Statistical Reasoning* (10% of questions)

Let's see how each one breaks down into more specific Test Day behaviors. Note that the bullet points of specific objectives for each of the SIRS are taken directly from the *Official Guide to the MCAT Exam*; the descriptions of what these behaviors mean and sample question stems, however, are written by Kaplan.

Skill 1: *Knowledge of Scientific Concepts and Principles*

This is probably the least surprising of the four SIRS; the testing of science knowledge is, after all, one of the signature qualities of the MCAT. Skill 1 questions will require you to do the following:

- Recognize correct scientific principles
- Identify the relationships among closely related concepts
- Identify the relationships between different representations of concepts (verbal, symbolic, graphic)
- Identify examples of observations that illustrate scientific principles
- Use mathematical equations to solve problems

At Kaplan, we simply call these Science Knowledge or Skill 1 questions. Another way to think of Skill 1 questions is as "one-step" problems. The single step is either to realize which scientific concept the question stem is suggesting or to take the concept stated in the question stem and identify which answer choice is an accurate application of it. Skill 1 questions are particularly prominent among discrete questions (those not associated with a passage). These questions are an opportunity to gain quick points on Test Day—if you know the science concept attached to the question, then that's it! On Test Day, 35% of the questions in each science section will be Skill 1 questions.

Here are some sample Skill 1 question stems:

- How would a proponent of the James–Lange theory of emotion interpret the findings of the study cited the passage?
- Which of the following most accurately describes the function of FSH in the human female menstrual cycle?
- If the products of Reaction 1 and Reaction 2 were combined in solution, the resulting reaction would form:
- Ionic bonds are maintained by which of the following forces?

Skill 2: *Scientific Reasoning and Problem-Solving*

The MCAT science sections do, of course, move beyond testing straightforward science knowledge; Skill 2 questions are the most common way in which it does so. At Kaplan, we also call these Critical Thinking questions. Skill 2 questions will require you to do the following:

- Reason about scientific principles, theories, and models
- Analyze and evaluate scientific explanations and predictions
- Evaluate arguments about causes and consequences
- Bring together theory, observations, and evidence to draw conclusions
- Recognize scientific findings that challenge or invalidate a scientific theory or model
- Determine and use scientific formulas to solve problems

Just as Skill 1 questions can be thought of as "one-step" problems, many Skill 2 questions are "two-step" problems, and more difficult Skill 2 questions may require three or more steps. These questions can require a wide spectrum of reasoning skills, including integration of multiple facts from a passage, combination of multiple science content areas, and prediction of an experiment's results. Skill 2 questions also tend to ask about science content without actually mentioning it by name. For example, a question might describe the results of one experiment and ask you to predict the results of a second experiment without actually telling you what underlying scientific principles are at work—part of the question's difficulty will be figuring out which principles to apply in order to get the correct answer. On Test Day, 45% of the questions in each science section will be Skill 2 questions.

Here are some sample Skill 2 question stems:

- Which of the following experimental conditions would most likely yield results similar to those in Figure 2?
- All of the following conclusions are supported by the information in the passage EXCEPT:
- The most likely cause of the anomalous results found by the experimenter is:
- An impact to a man's chest quickly reduces the volume of one of his lungs to 70% of its initial value while not allowing any air to escape from the man's mouth. By what percentage is the force of outward air pressure increased on a 2 cm^2 portion of the inner surface of the compressed lung?

Skill 3: *Reasoning About the Design and Execution of Research*

The MCAT is interested in your ability to critically appraise and analyze research, as this is an important day-to-day task of a physician. We call these questions Skill 3 or Experimental and Research Design questions for short. Skill 3 questions will require you to do the following:

- Identify the role of theory, past findings, and observations in scientific questioning
- Identify testable research questions and hypotheses
- Distinguish between samples and populations and distinguish results that support generalizations about populations
- Identify independent and dependent variables
- Reason about the features of research studies that suggest associations between variables or causal relationships between them (such as temporality and random assignment)
- Identify conclusions that are supported by research results
- Determine the implications of results for real-world situations
- Reason about ethical issues in scientific research

Over the years, the AAMC has received input from medical schools to require more practical research skills of MCAT test-takers, and Skill 3 questions are the response to these demands. This skill is unique in that the outside knowledge you need to answer Skill 3 questions is not taught in any one undergraduate course; instead, the research design principles needed to answer these questions are learned gradually throughout your science classes and especially through any laboratory work you have completed. It should be noted that Skill 3 comprises 10% of the questions in each science section on Test Day.

Here are some sample Skill 3 question stems:

- What is the dependent variable in the study described in the passage?
- The major flaw in the method used to measure disease susceptibility in Experiment 1 is:
- Which of the following procedures is most important for the experimenters to follow in order for their study to maintain a proper, randomized sample of research subjects?
- A researcher would like to test the hypothesis that individuals who move to an urban area during adulthood are more likely to own a car than are those who have lived in an urban area since birth. Which of the following studies would best test this hypothesis?

Skill 4: *Data-Based and Statistical Reasoning*

Lastly, the science sections of the MCAT test your ability to analyze the visual and numerical results of experiments and studies. We call these Data and Statistical Analysis questions. Skill 4 questions will require you to do the following:

- Use, analyze, and interpret data in figures, graphs, and tables
- Evaluate whether representations make sense for particular scientific observations and data
- Use measures of central tendency (mean, median, and mode) and measures of dispersion (range, interquartile range, and standard deviation) to describe data
- Reason about random and systematic error
- Reason about statistical significance and uncertainty (interpreting statistical significance levels and interpreting a confidence interval)
- Use data to explain relationships between variables or make predictions
- Use data to answer research questions and draw conclusions

Skill 4 is included in the MCAT because physicians and researchers spend much of their time examining the results of their own studies and the studies of others, and it's very important for them to make legitimate conclusions and sound judgments based on that data. The MCAT tests Skill 4 on all three science sections with graphical representations of data (charts and bar graphs) as well as numerical ones (tables, lists, and results summarized in sentence or paragraph form). On Test Day, 10% of the questions in each science section will be Skill 4 questions.

Here are some sample Skill 4 question stems:

- According to the information in the passage, there is an inverse correlation between:
- What conclusion is best supported by the findings displayed in Figure 2?
- A medical test for a rare type of heavy metal poisoning returns a positive result for 98% of affected individuals and 13% of unaffected individuals. Which of the following types of error is most prevalent in this test?
- If a fourth trial of Experiment 1 was run and yielded a result of 54% compliance, which of the following would be true?

SIRS Summary

Discussing the SIRS tested on the MCAT is a daunting prospect given that the very nature of the skills tends to make the conversation rather abstract. Nevertheless, with enough practice, you'll be able to identify each of the four skills quickly, and you'll also be able to apply the proper strategies to solve those problems on Test Day. If you need a quick reference to remind you of the four SIRS, these guidelines may help:

Skill 1 (Science Knowledge) questions ask:

- Do you remember this science content?

Skill 2 (Critical Thinking) questions ask:

- Do you remember this science content? And if you do, could you please apply it to this novel situation?
- Could you answer this question that cleverly combines multiple content areas at the same time?

Skill 3 (Experimental and Research Design) questions ask:

- Let's forget about the science content for a while. Could you give some insight into the experimental or research methods involved in this situation?

Skill 4 (Data and Statistical Analysis) questions ask:

- Let's forget about the science content for a while. Could you accurately read some graphs and tables for a moment? Could you make some conclusions or extrapolations based on the information presented?

CRITICAL ANALYSIS AND REASONING SKILLS (CARS)

The *Critical Analysis and Reasoning Skills* (CARS) section of the MCAT tests three discrete families of textual reasoning skills; each of these families requires a higher level of reasoning than the last. Those three skills are as follows:

1. *Foundations of Comprehension* (30% of questions)
2. *Reasoning Within the Text* (30% of questions)
3. *Reasoning Beyond the Text* (40% of questions)

These three skills are tested through nine humanities- and social sciences- themed passages, with approximately 5 to 7 questions per passage. Let's take a more in-depth look into these three skills. Again, the bullet points of specific objectives for each of the CARS are taken directly from the *Official Guide to the MCAT Exam*; the descriptions of what these behaviors mean and sample question stems, however, are written by Kaplan.

Foundations of Comprehension

Questions in this skill will ask for basic facts and simple inferences about the passage; the questions themselves will be similar to those seen on reading comprehension sections of other standardized exams like the SAT® and ACT®. *Foundations of Comprehension* questions will require you to do the following:

- Understand the basic components of the text
- Infer meaning from rhetorical devices, word choice, and text structure

This admittedly covers a wide range of potential question types including Main Idea, Detail, Function, and Definition-in-Context questions, but finding the correct answer to all *Foundations of Comprehension* questions will follow from a basic understanding of the passage and the point of view of its author (and occasionally that of other voices in the passage).

Here are some sample *Foundations of Comprehension* question stems:

- **Main Idea**—The author's primary purpose in this passage is:
- **Detail**—Based on the information in the second paragraph, which of the following is the most accurate summary of the opinion held by Schubert's critics?
- **(Scattered) Detail**—According to the passage, which of the following is FALSE about literary reviews in the 1920s?
- **Function**—The author's discussion of the effect of socioeconomic status on social mobility primarily serves which of the following functions?
- **Definition-in-Context**—The word "obscure" (paragraph 3), when used in reference to the historian's actions, most nearly means:

Reasoning Within the Text

While *Foundations of Comprehension* questions will usually depend on interpreting a single piece of information in the passage or understanding the passage as a whole, *Reasoning Within the Text* questions will typically require you to infer unstated parts of arguments or bring together two disparate pieces of the passage. *Reasoning Within the Text* questions will require you to:

- Integrate different components of the text to increase comprehension

In other words, questions in this skill often ask either *How do these two details relate to one another?* or *What else must be true that the author didn't say?* The CARS section will also ask you to judge certain parts of the passage or even judge the author. These questions, which fall under the *Reasoning Within the Text* skill, can ask you to identify authorial bias, evaluate the credibility of cited sources, determine the logical soundness of an argument, or search for relevant evidence in the passage to support a given conclusion. In all, this category includes Inference and Strengthen–Weaken (Within the Passage) questions, as well as a smattering of related—but rare—question types.

Here are some sample *Reasoning Within the Text* question stems:

- **Inference (Implication)**—Which of the following phrases, as used in the passage, is most suggestive that the author has a personal bias toward narrative records of history?
- **Inference (Assumption)**—In putting together her argument in the passage, the author most likely assumes:
- **Strengthen–Weaken (Within the Passage)**—Which of the following facts is used in the passage as the most prominent piece of evidence in favor of the author's conclusions?
- **Strengthen–Weaken (Within the Passage)**—Based on the role it plays in the author's argument, *The Possessed* can be considered:

Reasoning Beyond the Text

The distinguishing factor of *Reasoning Beyond the Text* questions is in the title of the skill: the word *Beyond*. Questions that test this skill, which make up a larger share of the CARS section than questions from either of the other two skills, will always introduce a completely new situation that was not present in the passage itself; these questions will ask you to determine how one influences the other. *Reasoning Beyond the Text* questions will require you to:

- Apply or extrapolate ideas from the passage to new contexts
- Assess the impact of introducing new factors, information, or conditions to ideas from the passage

The *Reasoning Beyond the Text* skill is further divided into Apply and Strengthen–Weaken (Beyond the Passage) questions, and a few other rarely appearing question types.

Here are some sample *Reasoning Beyond the Text* question stems:

- **Apply**—If a document were located that demonstrated Berlioz intended to include a chorus of at least 700 in his *Grande Messe des Mortes*, how would the author likely respond?
- **Apply**—Which of the following is the best example of a "virtuous rebellion," as it is defined in the passage?
- **Strengthen–Weaken (Beyond the Text)**—Suppose Jane Austen had written in a letter to her sister, "My strongest characters were those forced by circumstance to confront basic questions about the society in which they lived." What relevance would this have to the passage?
- **Strengthen–Weaken (Beyond the Text)**—Which of the following sentences, if added to the end of the passage, would most WEAKEN the author's conclusions in the last paragraph?

CARS Summary

Through the *Foundations of Comprehension* skill, the CARS section tests many of the reading skills you have been building on since grade school, albeit in the context of very challenging doctorate-level passages. But through the two other skills (*Reasoning Within the Text* and *Reasoning Beyond the Text*), the MCAT demands that you understand the deep structure of passages and the arguments within them at a very advanced level. And, of course, all of this is tested under very tight timing restrictions: only 102 seconds per question—and that doesn't even include the time spent reading the passages.

Here's a quick reference guide to the three CARS skills:

Foundations of Comprehension questions ask:

- Did you understand the passage and its main ideas?
- What does the passage have to say about this particular detail?

Reasoning Within the Text questions ask:

- What must be true that the author did not say?
- What's the logical relationship between these two ideas from the passage?
- How well argued is the author's thesis?

Reasoning Beyond the Text questions ask:

- How does this principle from the passage apply to this new situation?
- How does this new piece of information influence the arguments in the passage?

SCORING

Each of the four sections of the MCAT is scored between 118 and 132, with the median at 125. This means the total score ranges from 472 to 528, with the median at 500. Why such peculiar numbers? The AAMC stresses that this scale emphasizes the importance of the central portion of the score distribution, where most students score (around 125 per section, or 500 total), rather than putting undue focus on the high end of the scale.

Note that there is no wrong answer penalty on the MCAT, so you should select an answer for every question—even if it is only a guess.

The AAMC has released the 2017–2018 correlation between scaled score and percentile, as shown on the following page. It should be noted that the percentile scale is adjusted and renormalized over time and thus can shift slightly from year to year.

Total Score	Percentile	Total Score	Percentile
528	>99	499	47
527	>99	498	43
526	>99	497	40
525	>99	496	37
524	>99	495	33
523	>99	494	30
522	99	493	27
521	99	492	24
520	98	491	22
519	97	490	19
518	97	489	17
517	95	488	15
516	94	487	12
515	93	486	11
514	91	485	9
513	89	484	7
512	87	483	6
511	85	482	5
510	82	481	4
509	80	480	3
508	77	479	2
507	74	478	2
506	71	477	1
505	67	476	1
504	64	475	<1
503	61	474	<1
502	57	473	<1
501	54	472	<1
500	50		

Source: AAMC. 2018. *Summary of MCAT Total and Section Scores.* Accessed January 2018. **https://students-residents.aamc.org/advisors/article/percentile-ranks-for-the-mcat-exam/**.

Further information on score reporting is included at the end of the next section (see *After Your Test*).

MCAT POLICIES AND PROCEDURES

We strongly encourage you to download the latest copy of *MCAT® Essentials*, available on the AAMC's website, to ensure that you have the latest information about registration and Test Day policies and procedures; this document is updated annually. A brief summary of some of the most important rules is provided here.

MCAT Registration

The only way to register for the MCAT is online. You can access AAMC's registration system at: **www.aamc.org/mcat**.

You will be able to access the site approximately six months before Test Day. The AAMC designates three registration "Zones"—Gold, Silver, and Bronze. Registering during the Gold Zone (from the opening of registration until approximately one month before Test Day) provides the most flexibility and lowest test fees. The Silver Zone runs until approximately two to three weeks before Test Day and has less flexibility and higher fees; the Bronze Zone runs until approximately one to two weeks before Test Day and has the least flexibility and highest fees.

Fees and the Fee Assistance Program (FAP)

Payment for test registration must be made by MasterCard or VISA. As described earlier, the fees for registering for the MCAT—as well as rescheduling the exam or changing your testing center—increase as one approaches Test Day. In addition, it is not uncommon for test centers to fill up well in advance of the registration deadline. For these reasons, we recommend identifying your preferred Test Day as soon as possible and registering. There are ancillary benefits to having a set Test Day, as well: when you know the date you're working toward, you'll study harder and are less likely to keep pushing back the exam. The AAMC offers a Fee Assistance Program (FAP) for students with financial hardship to help reduce the cost of taking the MCAT, as well as for the American Medical College Application Service (AMCAS®) application. Further information on the FAP can be found at: **www.aamc.org/students/applying/fap**.

Testing Security

On Test Day, you will be required to present a qualifying form of ID. Generally, a current driver's license or United States passport will be sufficient (consult the AAMC website for the full list of qualifying criteria). When registering, take care to spell your first and last names (middle names, suffixes, and prefixes are not required and will not be verified on Test Day) precisely the same as they appear on this ID; failure to provide this ID at the test center or differences in spelling between your registration and ID will be considered a "no-show," and you will not receive a refund for the exam.

During Test Day registration other identity data collected may include: a digital palm vein scan, a Test Day photo, a digitization of your valid ID, and signatures. Some testing centers may use a metal detection wand to ensure that no prohibited items are brought into the testing room. Prohibited items include all electronic devices, including watches and timers, calculators, cell phones, and any and all forms of recording equipment; food, drinks (including water), and cigarettes or other smoking paraphernalia; hats and scarves (except for religious purposes); and books, notes, or other study materials. If you require a medical device, such as an insulin pump or pacemaker, you must apply for accommodated testing. During breaks, you are allowed to access food and drink, but not electronic devices, including cell phones.

Testing centers are under video surveillance and the AAMC does not take potential violations of testing security lightly. The bottom line: *know the rules and don't break them.*

Accommodations

Students with disabilities or medical conditions can apply for accommodated testing. Documentation of the disability or condition is required, and requests may take two months—or more—to be approved. For this reason, it is recommended that you begin the process of applying for accommodated testing as early as possible. More information on applying for accommodated testing can be found at: **www.aamc.org/students/applying/mcat/accommodations**.

After Your Test

When your MCAT is all over, no matter how you feel you did, be good to yourself when you leave the test center. Celebrate! Take a nap. Watch a movie. Ride your bike. Plan a trip. Call up all of your neglected friends or stalk them on Facebook. Totally consume a cheesesteak and drink dirty martinis at night (assuming you're over 21). Whatever you do, make sure that it has absolutely nothing to do with thinking too hard—you deserve some rest and relaxation.

Perhaps most importantly, do not discuss specific details about the test with anyone. For one, it is important to let go of the stress of Test Day, and reliving your exam only inhibits you from being able to do so. But more significantly, the Examinee Agreement you sign at the beginning of your exam specifically prohibits you from discussing or disclosing exam content. The AAMC is known to seek out individuals who violate this agreement and retains the right to prosecute these individuals at their discretion. This means that you should not, under any circumstances, discuss the exam in person or over the phone with other individuals—including us at Kaplan—or post information or questions about exam content to Facebook, Student Doctor Network, or other online social media. You are permitted to comment on your "general exam experience," including how you felt about the exam overall or an individual section, but this is a fine line. In summary: *if you're not certain whether you can discuss an aspect of the test or not, just don't do it!* Do not let a silly Facebook post stop you from becoming the doctor you deserve to be.

Scores are released approximately one month after Test Day. The release is staggered during the afternoon and evening, ending at 5 p.m. Eastern. This means that not all examinees receive their scores at exactly the same time. Your score report will include a scaled score for each section between 118 and 132, as well as your total combined score between 472 and 528. These scores are given as confidence intervals. For each section, the confidence interval is approximately the given score ±1; for the total score, it is approximately the given score ±2. You will also be given the corresponding percentile rank for each of these section scores and the total score.

AAMC CONTACT INFORMATION

For further questions, contact the MCAT team at the Association of American Medical Colleges:

MCAT Resource Center
Association of American Medical Colleges
www.aamc.org/mcat
(202) 828-0690
mcat@aamc.org

How This Book Was Created

The *Kaplan MCAT Review* project began shortly after the release of the *Preview Guide for the MCAT 2015 Exam*, 2nd edition. Through thorough analysis by our staff psychometricians, we were able to analyze the relative yield of the different topics on the MCAT, and we began constructing tables of contents for the books of the *Kaplan MCAT Review* series. A dedicated staff of 30 writers, 7 editors, and 32 proofreaders worked over 5,000 combined hours to produce these books. The format of the books was heavily influenced by weekly meetings with Kaplan's learning-science team.

In the years since this book was created, a number of opportunities for expansion and improvement have occurred. The current edition represents the culmination of the wisdom accumulated during that time frame, and it also includes several new features designed to improve the reading and learning experience in these texts.

These books were submitted for publication in April 2018. For any updates after this date, please visit www.kaptest.com/pages/retail-book-corrections-and-updates.

If you have any questions about the content presented here, email KaplanMCATfeedback@kaplan.com. For other questions not related to content, email booksupport@kaplan.com.

Each book has been vetted through at least ten rounds of review. To that end, the information presented in these books is true and accurate to the best of our knowledge. Still, your feedback helps us improve our prep materials. Please notify us of any inaccuracies or errors in the books by sending an email to KaplanMCATfeedback@kaplan.com.

Using This Book

Kaplan MCAT Critical Analysis and Reasoning Skills Review, and the other six books in the *Kaplan MCAT Review* series, bring the Kaplan classroom experience to you—right in your home, at your convenience. This book offers the same Kaplan content review, strategies, and practice that make Kaplan the #1 choice for MCAT prep.

This book is designed to help you review the *Critical Analysis and Reasoning Skills* section of the MCAT. Unlike other books in this MCAT series, there is no content to review for the *Critical Analysis and Reasoning Skills* section. The questions are written in such a way that they do not presume any prior fund of knowledge. In other words, all the support that is needed to answer the questions correctly is found in the corresponding pages.

LEARNING GOALS

At the beginning of each chapter, you'll find a short list of objectives describing the skills covered within that chapter. Learning goals for these texts were developed in conjunction with Kaplan's learning science team, and have been designed specifically to focus your attention on tasks and concepts that are likely to be relevant to your MCAT testing experience. These learning goals will function as a means to guide your review of the chapter, and indicate what information and relationships you should be focused on within each chapter. Before starting each chapter, read these learning goals carefully. They will not only allow you to assess your existing familiarity with the content of the chapter, but also provide a goal-oriented focus for your studying experience.

SIDEBARS

The following is a guide to the five types of sidebars you'll find in *Kaplan MCAT Critical Analysis and Reasoning Skills Review*:

- **Bridge:** These sidebars create connections between science topics that appear in multiple chapters throughout the *Kaplan MCAT Review* series.
- **Key Concept:** These sidebars draw attention to the most important takeaways in a given topic, and they sometimes offer synopses or overviews of complex information. If you understand nothing else, make sure you grasp the Key Concepts for any given subject.
- **MCAT Expertise:** These sidebars point out how information may be tested on the MCAT or offer key strategy points and test-taking tips that you should apply on Test Day.
- **Mnemonic:** These sidebars present memory devices to help recall certain facts.
- **Real World:** These sidebars illustrate how a concept in the text relates to the practice of medicine or the world at large. While this is not information you need to know for Test Day, many of the topics in Real World sidebars are excellent examples of how a concept may appear in a passage or discrete (stand-alone) question on the MCAT.

In the end, this is your book, so write in the margins, draw diagrams, highlight the key points—do whatever is necessary to help you get that higher score. We look forward to working with you as you achieve your dreams and become the doctor you deserve to be!

ONLINE RESOURCES

In addition to the resources located within this text, you also have additional online resources awaiting you at **www.kaptest.com/booksonline**. Make sure to log on and take advantage of free practice and access to online versions of the book!

Please note that access to the online resources is limited to the original owner of this book.

About CARS

1

1: About CARS

In This Chapter

1.1 The CARS Section	4	Concept and Strategy Summary	8
1.2 Passages	4		
1.3 Question Categories	5		
Foundations of Comprehension	5		
Reasoning Within the Text	6		
Reasoning Beyond the Text	6		

Introduction

LEARNING GOALS

After Chapter 1, you will be able to:

- Recite the major structural features of the CARS section of the MCAT
- Recall the two major passage topic categories
- Explain the major differences between *Foundations of Comprehension*, *Reasoning Within the Text*, and *Reasoning Beyond the Text* question categories

Congratulations! You are about to embark upon an exciting journey down the path to medical school to achieve your goal of becoming a doctor. Like any major journey in life, this will require thorough preparation. Fortunately, you don't have to prepare for this journey on your own: Kaplan offers comprehensive preparation for the MCAT with proven strategies that can enhance your score in all sections of the exam.

As a premedical student, you have already been exposed to a wide variety of science topics that will be tested in the three science sections of the exam. In contrast, the *Critical Analysis and Reasoning Skills* (CARS) section will present you with an array of passages from various disciplines to which you may have never been exposed—for example, a musicological analysis of Johannes Brahms's 1868 masterpiece *Ein Deutsches Requiem*, a philosophical diatribe criticizing Immanuel Kant's *Metaphysics*, or a dissection of the political underpinnings of the development of the Medicare system—and you will be expected to read, understand, and apply this knowledge. Students often find themselves overwhelmed or feel ill-equipped for the CARS section of the test, but Kaplan is here to help! This book will help you understand what is expected of you in CARS and presents the Kaplan strategies that have paved the way for many thousands of students to become the doctors they deserve to be.

MCAT Critical Analysis and Reasoning Skills

In this chapter, we will go over the structure of the CARS section of the MCAT, as well as the diverse disciplines encountered in CARS passages. We'll provide a brief overview of the question categories identified by the Association of American Medical Colleges (AAMC). Finally, we'll discuss how to use this book and how it can guide you in preparing for your MCAT and the journey beyond. The journey to becoming a physician may be long, but it is ultimately extremely rewarding. Some day in the future, you'll find yourself donning your white coat, changing patients' lives, and realizing that having the right plan for success is what made the journey possible.

1.1 The CARS Section

In some ways, the *Critical Analysis and Reasoning Skills* (CARS) section of the MCAT will be nothing new to you; it is similar to many of the standardized tests you have taken throughout your academic career, presenting you with passages to read and multiple-choice questions to gauge your understanding. In 90 minutes, you will be presented with 9 passages, each of which will be followed by approximately 5 to 7 questions, for a total of 53 questions. The passages you encounter will be relatively short (but lengthier than the science passages on the test), ranging from 500 to 600 words.

Unlike reading comprehension sections you have come across previously, such as those in the SAT® or ACT®, the CARS section of the MCAT has been designed to assess analytical and reasoning skills that are required in medical school. The passages you will face in CARS will be multifaceted, incorporating advanced vocabulary, presenting varied writing styles, and requiring higher-level thought. To answer the accompanying questions, you will have to go beyond merely comprehending the content of a CARS passage: you will need to analyze its rhetorical and logical structure, and even be able to assess how it impacts (or is impacted by) outside information.

1.2 Passages

The types of passages chosen for CARS consist of multiple paragraphs that require active, critical reading to answer the questions that follow. The passages included in the section are from an array of disciplines in the social sciences and humanities, as listed in Table 1.1. Approximately half of the passages (and questions) that you encounter on Test Day will fall in the realm of the humanities, while the other half will be in the social sciences. All of the passages that appear in CARS are selected from books, journals, and other publications similar to those you have come across in academic settings.

Humanities	Social Sciences
Architecture	Anthropology
Art	Archaeology
Dance	Economics
Ethics	Education
Literature	Geography
Music	History
Philosophy	Linguistics
Popular Culture	Political Science
Religion	Population Health
Studies of Diverse Cultures*	Psychology
Theater	Sociology
	Studies of Diverse Cultures*

* Note: Studies of Diverse Cultures can be tested in both humanities and social sciences passages.

Table 1.1. Humanities and Social Sciences Disciplines in the CARS Section[1]

For students who have exclusively focused on the sciences, information for the fields used in the CARS section may be presented in a strikingly different way that can sometimes seem overwhelming. This book will review the writing styles used for the passages in CARS and explain how to read these passages with purpose, which will ultimately make them much less intimidating and significantly more manageable.

1.3 Question Categories

The AAMC has identified three categories of questions in CARS that will assess your critical thinking skills: *Foundations of Comprehension, Reasoning Within the Text*, and *Reasoning Beyond the Text*.

FOUNDATIONS OF COMPREHENSION

These questions tend to be straightforward. They will ask about the main ideas of a passage, specific details from within the passage, the purpose of a given part of the passage, or the likely meaning of a word or phrase based on context. These questions are the most similar to those you have seen in previous standardized tests because they ask only for reading comprehension (understanding what you have read). Questions in *Foundations of Comprehension* will make up approximately 30 percent of the questions in CARS, or 16 questions.

---

1. AAMC, *The Official Guide to the MCAT 2015 Exam* (Washington, D.C.: Association of American Medical Colleges, 2014), 311–22.

MCAT Critical Analysis and Reasoning Skills

In Chapter 9 of *MCAT CARS Review*, we will further dissect the four question types within *Foundations of Comprehension*:

- Main Idea
- Detail
- Function
- Definition-in-Context

REASONING WITHIN THE TEXT

Reasoning Within the Text questions require greater thought than *Foundations of Comprehension* questions because they will ask you to draw inferences (unstated parts of arguments that logically must be true based on the information given) or ask how one piece of information relates to another (as a piece of evidence that supports a conclusion, for example). Questions in *Reasoning Within the Text* will also make up approximately 30 percent of the questions in CARS, or 16 questions.

In Chapter 10 of *MCAT CARS Review*, we will further dissect the two main question types within *Reasoning Within the Text* and a few other, rare questions that fit into this category:

- Inference
- Strengthen–Weaken (Within the Passage)

REASONING BEYOND THE TEXT

Reasoning Beyond the Text questions focus on two specific skills: first, the capacity to extrapolate information from the passage and place it within new contexts and, second, the ability to ascertain how new information would relate to and affect the concepts in the passage. Questions in *Reasoning Beyond the Text* will make up approximately 40 percent of the questions in CARS, or 21 questions.

In Chapter 11 of *MCAT CARS Review*, we will further dissect the two main question types within *Reasoning Beyond the Text* and a few other, rare questions that fit into this category:

- Apply
- Strengthen–Weaken (Beyond the Passage)

Conclusion

This chapter is only a beginning. Now that we have covered the structure of the CARS section, we will dive into the Kaplan strategies that will help you score points on Test Day. In Chapter 2, we will begin with an analysis of rhetoric, exploring how a passage's author creates a conduit for the transmission of a message to an audience.

In Chapter 3, we will discuss critical reading and, specifically, how keywords will underlie your ability to master the four modes of reading. We continue that conversation in Chapter 4 through the Kaplan Method for CARS Passages. Chapters 5 and 6 similarly constitute a single unit; in Chapter 5, we look at how arguments are constructed, and in Chapter 6, we bring in introductory formal logic to create a systematized approach to these arguments. Chapter 7 serves as a turning point as we segue from discussing the varieties of passages and common passage structures to anticipating questions. In Chapter 8, we introduce the Kaplan Method for CARS Questions, and then we see its application in the three following chapters to each of the different AAMC categories. Finally, we end with a look at how to review your practice tests to find your personal test-taking pathologies and keep improving that score.

MCAT Critical Analysis and Reasoning Skills

CONCEPT AND STRATEGY SUMMARY

The CARS Section

- The *Critical Analysis and Reasoning Skills* (CARS) section lasts 90 minutes and contains 53 questions, divided among 9 passages.
  - Passages range from 500 to 600 words.
  - Each passage has approximately 5 to 7 questions.
- CARS requires going beyond merely comprehending the content: you must analyze a passage's rhetorical and logical structure and even be able to assess relationships between information given in the passage and new outside information.

Passages

- Half of the passages will be in the humanities (architecture, art, dance, ethics, literature, music, philosophy, popular culture, religion, studies of diverse cultures, and theater).
- Half of the passages will be in the social sciences (anthropology, archaeology, economics, education, geography, history, linguistics, political science, population health, psychology, sociology, and studies of diverse cultures).

Question Categories

- *Foundations of Comprehension* questions ask about the main ideas of a passage, specific details from within the passage, the purpose of a given part of the passage, or the likely meaning of a word or phrase based on context.
- *Reasoning Within the Text* questions ask you to draw inferences (unstated parts of arguments that logically must be true based on the information given) or ask how one piece of information relates to another (as a piece of evidence that supports a conclusion, for example).
- *Reasoning Beyond the Text* questions ask you to extrapolate information from the passage and place it within a new context or to ascertain how new information would relate to and affect the concepts in the passage.

2
Analyzing Rhetoric

2: Analyzing Rhetoric

In This Chapter

2.1 What Is Rhetoric?	12	2.3 Rhetorical Analysis	17
2.2 Key Components of Rhetoric	12	Examples	17
Author	13	Aristotelian Rhetoric	20
Audience	13	**Concept and Strategy Summary**	**22**
Message	14		
Goal	15		
Context	16		

Introduction

> **LEARNING GOALS**
>
> After Chapter 2, you will be able to:
>
> - Explain the importance of rhetorical analysis in approaching the CARS section
> - Describe the key components of a rhetorical situation: author, audience, message, goal, and context
> - Apply principles of rhetorical analysis to MCAT-style passages
> - Recall the role of logos, ethos, and pathos in Aristotelian rhetoric

You turn on the television and see a grainy, black-and-white, unflattering portrait of a woman with disconcerting statistics displayed to her right, ominous music in the background, and a narrator who sounds apprehensive as he recites all of her apparent flaws. *Another one of those stupid political attack ads*, you think as you flip the channel. Suddenly you find yourself confronted with startling images of malnourished children surrounded by filth as you hear a woman talking about *the importance of your donation*. You immediately recognize this as a commercial for a charitable organization and quietly change channels again to avoid feeling guilty. Now, an educated man standing behind a podium appears on the screen, droning on in a monotone about why his interpretation of the causes of the War of 1812 is the correct one. *Must be one of those public-access educational channels*, you conclude. *Maybe I should just get back to studying for my MCAT*. Though you may not realize it as you put down your remote control, simply by making conclusions about what you were watching on TV, you were engaged in rhetorical analysis.

MCAT Critical Analysis and Reasoning Skills

In this chapter, we will take a more systematic approach to analyzing rhetoric. First, we will define the terms used in rhetoric and consider how rhetoric plays a role in daily life. Then, we will demonstrate the application of this knowledge to the MCAT. As you read, you will come to see the role rhetoric has already played in your life as a student and the value of increasing your awareness of it as you study for the CARS section. Why are rhetorical skills tested on the MCAT? Ultimately, a keen awareness of the topics discussed in this chapter will help you think critically about *how* information is delivered in medical school and beyond.

2.1 What Is Rhetoric?

Most of us are familiar with the device known as a rhetorical question. Although it ends in a question mark, a rhetorical question tends to have only one plausible and obvious answer. What makes it rhetorically effective is that it forces readers to reach the conclusion themselves, so they are more convinced of it than if the author had simply stated it. However, there are many more nuances to rhetoric than questions that aren't really questions. Considered broadly, **rhetoric** is the art of effective communication, both in speech and in text. Because the MCAT is a written exam, we will predominantly discuss the textual side of rhetoric throughout this chapter. While language may serve many purposes, the study of rhetoric tends to focus on persuasion—the attempt to influence others to adopt particular beliefs or engage in certain behaviors. **Rhetorical analysis**, then, is an examination of speech or writing that goes beyond *what* the author is saying (the content) to consider *how* the author is saying it, with a particular emphasis on techniques of persuasion.

When the *Writing Sample* was a section on the MCAT (from 1991 to 2013), effective *use* of rhetoric was an important skill to hone for Test Day. While premedical students still need to be able to write clearly for the Personal Statement and other application essays, your principal concern in the CARS section will be to understand how *other* people use rhetoric, in other words, to use rhetorical analysis. To that end, we will begin by defining several of the fundamental aspects of rhetoric: author, audience, message, goal, and context. We will then discuss some more subtle elements of rhetoric. Together, these constitute what we'll call **rhetorical knowledge**—an awareness of the other aspects of a text besides its content—which is necessary for the process of rhetorical analysis.

2.2 Key Components of Rhetoric

The **rhetorical situation** is a way of representing any act of communication, emphasizing the transmission of ideas from an individual to an audience. Effective authors

> **Key Concept**
> The rhetorical situation focuses on the transmission of ideas from the author to the audience, with a particular goal in mind.

are rhetorically savvy and will direct their message to a particular subset of people with a clear goal in mind. Before placing pen to paper, or tapping out thoughts on the keyboard, the writer must answer the questions *To whom am I writing?* and *Why?*

AUTHOR

The **author**, in the most basic sense, is the individual or group writing the text. Authors can be distinguished by how much expertise they have on the topic at hand, by how passionate or vested in the topic they are, and by the groups or stakeholders they represent.

Authors who are experts in a topic—and who know that their intended audiences are also knowledgeable in the topic—tend to use a lot of jargon in their writing. **Jargon** refers to technical words and phrases that belong to a particular field. For example, *transcriptional repression*, *zwitterion*, and *anabolism* are all biochemical jargon; *homunculus*, *Gesellschaften*, and *negative symptoms* are all behavioral sciences jargon. Authors who are less expert, or who are writing to a less-informed audience, tend to use more common terminology and provide more explicitly detailed descriptions of their ideas. Authors who consider themselves less expert than their audiences may use an abundance of Moderating keywords, described in Chapter 3 of *MCAT CARS Review*.

> **MCAT Expertise**
>
> The more knowledgeable authors are in a topic, the more jargon may appear in their writing. This may make for a challenging passage to read, but recognize the MCAT does not expect you to know any field-specific terminology in CARS. Any important jargon will be defined in the passage—or the definition will be strongly implied.

When an author is passionate about a topic, this emotion often manifests as strong language. Extreme keywords, also described in Chapter 3 of *MCAT CARS Review*, may suggest that an author is emotionally invested in the piece. Less-invested authors may use more emotionally neutral words to describe the same ideas.

Finally, an author may be writing on behalf of a larger group or body of stakeholders—individuals who also have an interest in the outcome of the written piece. This situation is sometimes indicated by an author's use of the word *we* rather than *I*. When speaking for others, authors may strengthen or weaken the representation of their opinions in the piece to match those of the rest of the group.

AUDIENCE

The **audience** is the person, or persons, for whom the text is intended. In daily life, the audience could be a single person with whom you have a dialogue, but publications typically have considerably larger audiences. Many CARS passages address an academic audience—perhaps other specialists in the author's field. Even when writing for the "general public," authors will draw upon idioms, clichés, symbols, and references that are recognizable only to people of a particular time and place. Because of the wide availability of many academic journals and historical documents, the author's message may reach a much wider audience than the author originally intended.

MCAT Critical Analysis and Reasoning Skills

Regardless of whom the piece is written for, each person who reads it will approach the text differently. Nevertheless, readers who share many characteristics will tend to interpret the work in similar ways. For instance, students preparing to take the MCAT tend to share a science background that people who are interested in pursuing a business degree may lack. A business student might respond well to an article on the profitability of an industrial chemical process, but the same reader would likely need additional background to read a piece that proposed a theoretical mechanism explaining the chemical reaction's kinetics. An MCAT student, on the other hand, may have more difficulty navigating complex business terminology but would certainly be able to understand the mechanism description after reading Chapter 5 of *MCAT General Chemistry Review*. In CARS, you will rarely be a member of the intended audience, but you can still develop the ability to recognize for whom a passage was originally written.

MESSAGE

The **message** is the actual text that the author writes. This may be a physical document—as in the case of a journal article, book, or email—or it may be delivered verbally. It is important to note that the message itself contains only the explicit information offered by the author—the facts, data, and concepts the author draws upon, as well as any stated opinions.

The CARS section is unique in that many of its questions are not solely focused on the message the author has written but instead emphasize the goal and context of the piece. Whereas many other standardized tests focus predominantly on understanding what the author has said, the MCAT takes it a step further, sometimes asking you to use the text to glean characteristics about the author (including biases, assumptions, and even identity markers such as profession).

Voice

The author's **voice** is her unique style while writing. The author's word choice and the way she constructs sentences will differ from that of any other author, and may offer hints about her personality or other aspects of her identity. For example, you may be able to identify which of your friends wrote a particular email based solely on its wording. In CARS passages, quotations from other writers will present a different voice than that used by the author of the passage.

Genre and Medium

The **genre** is the category the written work belongs to. Genre encompasses a wide array of classifications, including fiction, nonfiction, drama, poetry, and so on. Genre can also represent the more concrete form of the work: a book, scholarly journal article, case study, essay, letter, email, and so forth. The audience and goal will necessarily affect what genre the author uses for the piece.

Bridge

An author's voice is related to two components of language theory: syntax (word choice and word order) and pragmatics (adapting one's message depending on the social context). Language theory is discussed in Chapter 4 of *MCAT Behavioral Sciences Review*.

Real World

Why do the testmakers care so much about your ability to think beyond just the message of an author's writing? This is a skill you'll use every day when talking with patients (to discern what they are really thinking from their body language and tone of voice) and when critically appraising research (to look for potential biases or conflicts of interest on the part of the researcher). To make good choices for your patients, you must be able to move beyond simply reading words on a page (message) and consider the bigger picture (goal and context).

The **medium** is the delivery system into which the written work can be placed. In other words, it is the method used to transfer the message from the author to the audience. Examples of media include print (such as books, academic journals, newspapers, or pamphlets), broadcast (television, radio, music), and digital (email, text messages, social networking).

GOAL

The **goal** is the author's intended outcome—the effect that she wishes to produce with her writing. In some cases, the author's goal may be simply to inform the audience. Passages with an informative goal tend to read like textbooks or encyclopedia entries, providing detailed descriptions nearly devoid of the author's opinion. More often the goal is persuasion, in which the author aims to influence the audience to adopt new beliefs. Although some authors seek to recount facts impartially, even informing the audience is a kind of persuasion insofar as it causes readers to form new opinions. Persuasive passages are the most common on the MCAT; most passages on Test Day will contain at least one opinion the author tries to get the reader to endorse (with varying degrees of forcefulness). Besides altering beliefs, persuasion can also motivate individuals to take action. Such persuasion is often encountered in speeches, but it could appear on the MCAT as a set of recommendations for solving a particular problem.

Evoking an emotional response is another kind of goal that may be distinct from persuasion. Whether the author seeks to delight the audience with humorous anecdotes or to provoke curiosity by raising unanswered questions, such a goal may be achieved by dramatically different means from those used in persuasive writing. These two categories need not be separate—certainly, an author can evoke an emotional response in an attempt to persuade the audience.

More generally, an author's tone helps to establish the goal of a written work. **Tone** reflects the author's attitude toward the subject matter. In a literary piece, the way in which the author frames the characters in the text—their personalities, dialogues, and actions—contributes to the tone. In an article, how the author describes the scenario and the individuals or entities being discussed reveals the tone. To determine the tone of a text, examine the words that are used while considering the question *What imagery or feelings do these words convey?*

Whether a piece is persuasive, is evocative, or serves another goal, it will invariably attempt to affect the members of an audience in some way by changing what they feel, what they believe, or what they do. Bear in mind that just as the intended audience can be misjudged, so too can the actual effect of a text depart from its author's original desires.

> **Bridge**
> Impression management, discussed in Chapter 9 of *MCAT Behavioral Sciences Review*, focuses on how we present ourselves to accomplish specific goals. Some of the impression management strategies can be employed in writing as well, such as self-disclosure, ingratiation, and altercasting.

> **Bridge**
> An author's Goal should be placed at the bottom of your Outline, and part of writing out that Goal is identifying the tone: is the author prescriptive? Demanding? Doubtful? Impartial? How to Outline a passage is described in Chapter 4 of *MCAT CARS Review*.

MCAT Critical Analysis and Reasoning Skills

CONTEXT

The **context** of a written work refers to two distinct—but related—aspects of the work: the relationships among words, sentences, and paragraphs; and the larger societal situation in which the piece was written.

Within a written work, the term *context* often refers to surrounding information. For example, the definition or intent of a word can often be discerned by reading the surrounding sentence. This approach to context can be used on a larger scale as well. The intent of a sentence can become clearer when you read the surrounding sentences, and the significance of a paragraph is dependent on the paragraphs that precede and follow it. Even a full chapter can be understood more clearly in relation to a book as a whole. Context helps with placing the author's ideas into perspective.

The term *context* can also refer to the societal situation in which the author wrote the piece. Recognizing the author, audience, and message of a piece (as well as the relationships between each) leads to a better understanding of the goal of the work. This means not only literally identifying each of these elements of the rhetorical situation but also looking beyond the pages of the text to the social, political, economic, cultural, and intellectual environment at the time. For example, *I Have a Dream* by Dr. Martin Luther King, Jr. is far better understood when one considers the impetus for the civil rights movement in the early 1960s than if one reads the words without any awareness of their historical context. While the MCAT will not require you to know anything about the sociohistorical context outside of the information contained in the passage (or in the accompanying question stems), it is critical to recognize that the author's tone and voice—further illustrated in the next section—are influenced by context in this broader sense.

Let's return to why the rhetorical situation, diagrammed in Figure 2.1, is so important in medicine. Many people are interested in eating healthfully and staying fit, and many people are interested in expanding their knowledge about their bodies. However, most patients with whom you interact will not be reading the *Journal of the American Medical Association* (*JAMA*) or the *New England Journal of Medicine* (*NEJM*), which routinely publish medical studies with this information. Instead, most will find information in health and wellness magazines, in online forums, or on popular television shows. These media deliver this health information in a more understandable way for those who are not highly educated in medicine.

Medical journals and the more popular media are both intended to inform; however, one is aimed at vetting the latest research through peer review and delivering that research to health professionals, while the other presents conclusions and recommendations to the general public. The difference between these two rhetorical situations can lead to vastly different interpretations of the same research.

> **Bridge**
>
> Definition-in-Context questions, discussed in Chapter 9 of *MCAT CARS Review*, are all about your ability to ascertain the definition of a word (usually jargon) from its surrounding text.

2: Analyzing Rhetoric

A less-informed blogger may present faulty conclusions, a magazine partially funded by pharmaceutical companies may be biased in its descriptions of a particular medication, and a sensationalist television show may exaggerate the importance of a discovery. By recognizing the rhetorical situation in which information is presented, physicians can better understand how their patients think and can advise and work with these patients more effectively.

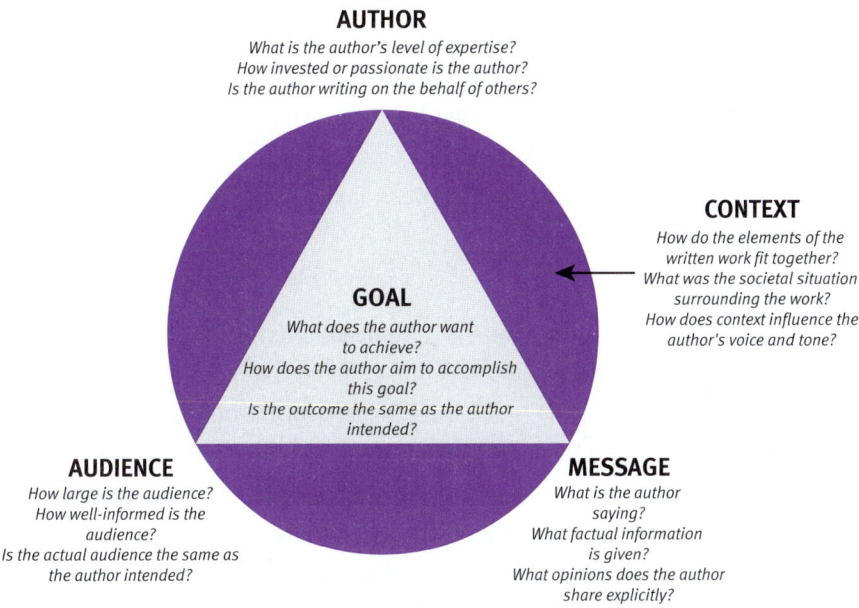

Figure 2.1. The Rhetorical Situation

2.3 Rhetorical Analysis

Now that we have established a common vocabulary to discuss rhetoric, let's take a look at rhetorical analysis—the application of rhetorical knowledge.

EXAMPLES

The following are three contrasting passages that demonstrate the conventions of rhetorical knowledge. All three of the passages focus on the same topic: Aristotle's *Rhetoric*.

Example Passage 1

Aristotle's brilliance is further on display in the study of rhetoric. The canonical *Rhetoric*, much in the fashion of his numerous other groundbreaking treatises, lays the foundation for an entire academic discipline, elucidating rhetoric's multifarious nature as Aristotle employs a series of clarifying distinctions. Thus, Aristotle demonstrates the vast scope of

> **MCAT Expertise**
>
> When you see examples like these three passages in this book, try to come up with answers on your own before looking at Kaplan's answers. In this set of examples, determine the message, voice, goal, and tone of each example. If you train yourself to think critically while you practice, it will be second nature to do so on Test Day.

rhetoric by calling it the *antistrophe* or counterpart to *dialectic*. While the dialectician attempts to determine what is actually true or ethically good, the rhetorician tries to persuade others of the truth or goodness of something. Among his most remarkable insights is Aristotle's division of the three modes of persuasion. When orators draw upon rational argumentation to convince an audience, they are employing *logos*, or logical persuasion. If speakers wish to establish credibility to demonstrate that they know what they're talking about and that they are not being deceptive, they use *ethos*, or the ethical mode. Finally, *pathos*, or emotional persuasion, is utilized whenever a speaker attempts to influence the audience's feelings to incline its members to think or act in a particular way. Of course, Aristotle in his wisdom recognizes that effective oratory almost invariably displays characteristics of all three.

Example Passage 2

Another area in which Aristotle was influential is the field of rhetoric. In his book entitled *Rhetoric*, Aristotle employs his typical approach of categorization, delineating the various aspects of the subject of inquiry. According to Aristotle, rhetoric is the practical complement to the more theoretical *dialectic*: the latter concerns matters of truth and goodness while the former concerns the means of persuasion. This persuasion is accomplished by using one or more of three basic modes or methods: *logos*, *ethos*, and *pathos*. *Logos*, or logical persuasion, relies upon reason, employing both deductive proofs (*syllogisms* or *enthymemes*) and inductive inferences made through the use of examples. *Ethos*, or ethical persuasion, aims to draw upon the credibility of the speaker to establish the speaker's expertise and goodwill. Lastly, *pathos*, or pathetic persuasion, sways individuals by playing upon their emotions, putting them in the proper state of mind to form beliefs or take actions as the speaker wills. Most orators will use all three methods of persuasion but will frequently emphasize one over the others.

Example Passage 3

Despite the advances of later thinkers, Aristotle continues to have a significant impact on rhetorical studies. His *Rhetoric* is all too often cited as providing a foundation for the entire discipline, notwithstanding Aristotle's penchant for simplistic classifications. One example of such is his makeshift division of all academic disciplines into two categories: theoretical *dialectic* and practical *rhetoric*. Aristotle contends that the latter operates in only three distinct ways: *logos*, the use of reason to demonstrate to an audience the truths that dialectic purportedly uncovers; *ethos*, persuasion that exploits the speaker's apparent trustworthiness and expertise; and *pathos*, or emotional manipulation of the audience. Fortunately, Aristotle does not deny common sense, granting that most persuasion is a mixture of the three—although one wonders why he does not leave open the possibility of alternative methods.

There are commonalities in the passages. All three are written in an articulate manner using advanced vocabulary, which hints that the authors are highly knowledgeable about Aristotelian classifications of rhetoric. The advanced vocabulary also signals that the passages are designed for educated audiences.

Before reading the interpretations of the three passages below, decide for yourself how the principles of rhetoric apply to each example. What message does each author give? How easy is it to discern the author's voice? What is the author's goal, and how does the tone help achieve that goal?

In the first passage, the author shows that he favors Aristotle by painting the philosopher in a positive manner: *Aristotle's brilliance*, *Among his most remarkable insights*, and *Aristotle in his wisdom* are a few examples of language through which the author reveals a positive bias toward Aristotle. While the author of the first passage explains Aristotle's philosophy to the audience, the message also includes the author's subjective views. The author's voice is clearly evident, and the tone is light, respectful, and complimentary. This type of passage would likely be found in an arts and culture publication in which the goal is to highlight or celebrate the philosopher's contributions to society.

The second passage is devoid of the author's personal views and presents the information in an objective manner. That is, the message of the second example is more purely focused on providing accurate information about the Aristotelian classifications of rhetoric without including explicit opinion. The tone is more serious and strictly informative, and the author seems to have minimized her personal voice—or has a very dry and factual voice. This type of text may appear in an encyclopedia or perhaps a textbook in which the goal is simply to provide facts.

In contrast, the final passage is rather dismissive: *Despite the advances of later thinkers, Aristotle continues to have a significant impact*; *too often cited*; *Aristotle's penchant for simplistic classifications*; and *Fortunately, Aristotle does not deny common sense* all reveal a clear negative bias against Aristotle. The author trivializes the contributions made by Aristotle and questions the philosopher's ideology. Such text could be an excerpt from an article in an academic journal in which the author was focusing on one of Aristotle's successors or providing criticisms of Aristotle. Either way, the tone is critical, as the author's opposition is evident; the author's voice is also obvious, as in the first example. The author's goal is unmistakably to downplay or rebut Aristotle's importance while describing the classifications in *Rhetoric*.

MCAT Critical Analysis and Reasoning Skills

ARISTOTELIAN RHETORIC

We would be remiss if we did not provide our own definitions of Aristotle's rhetorical classifications after the last set of examples. According to the Greek philosopher, rhetoric relies on the communication skills of both the author of the message and the audience receiving the message. Three strategies, also called **appeals**, are used in this approach: *logos*, *ethos*, and *pathos*.

The **Aristotelian triad** has some parallels to the rhetorical triangle of author, audience, and message illustrated in Figure 2.1. For effective transmission, the message conveyed by the author needs to be logically sound and based firmly in reason. Aristotle classified ***logos*** as an appeal to one's logic and reasoning. The author also needs to be credible and must be presented as an ethical deliverer of the message. Creating an appeal via one's credibility and trustworthiness is termed ***ethos***. Lastly, the author is often more successful by appealing to the audience's emotions; such an appeal is termed ***pathos***. While it is the author's role to establish the logos, ethos, and pathos of a message, the audience members are also expected to use their critical reading or listening skills and reasoning abilities to analyze the message conveyed by the author.

When reading MCAT passages, it is important to consider the frame of mind of the author and her intent in writing the passage. This perspective gives insight into how an author uses rhetorical strategies to appeal to the audience. *Logos* in particular is especially question-worthy, so logical reasoning is addressed further in Chapters 5 and 6 of *MCAT CARS Review*.

> **MCAT Expertise**
>
> The MCAT will not ask you to identify which of the Aristotelian triad of appeals is used by an author, but the test will certainly ask you to identify *how* an author supports his argument in the passage.

Conclusion

All of the questions in CARS are associated with passages that you must read critically. But, as described earlier, most questions will go beyond simply comprehending the words the author says (the message) and will instead focus on higher-level concerns. What does the author's tone reveal about her opinions? How does the author's voice reflect her identity and intended audience? What is the author's goal in writing the passage? Questions like these, which reflect the underlying principles of rhetoric are often encountered on Test Day.

Of course, rhetoric is not just an antiquated discipline passed down from the ancient Greeks, nor is rhetorical analysis a skill you will apply only on the MCAT. You will use these same ideas and techniques on a daily basis in medical school and as a physician. In medical school, you will learn not only the language of medicine but also the rhetorical devices used by physicians—that is, how to author a patient note that seamlessly guides other physicians to the same diagnosis as you. *A 75-year-old woman with a history of Factor V Leiden presents on post-operative day #5 status post right hip replacement with swelling of the right leg, acute onset of shortness of breath, and chest pain*—from this one sentence, a doctor is able to communicate the same message to any other physician who reads the note: *I think this woman has a deep venous thrombosis (DVT), which has sent a piece of the clot to her lungs.*

With the expansion of evidence-based medicine, you'll answer clinical questions through a meticulous analysis of the research on the topic: *Which breast cancer screening guidelines should I follow?* you may ask yourself. *Do I opt for the United States Preventive Services Task Force (USPSTF), which tends to be more conservative with screening, or the American Cancer Society (ACS), which is more rigorous? Or what about the American Congress of Obstetricians and Gynecologists (ACOG), which is somewhere in between?* An effective physician would read each group's recommendations with an eye toward the identity of each group (the authors) and each group's goals. Examples such as these make it clear that understanding rhetoric is indispensable to success on the MCAT, in medical school, and beyond.

CONCEPT AND STRATEGY SUMMARY

What Is Rhetoric?
- **Rhetoric** is the art of communicating through writing and speaking.
- **Rhetorical knowledge** refers to an awareness of the components of a written work besides the actual words on the page, such as the author, intended audience, and goal.
- **Rhetorical analysis** is the examination of a particular work for the sake of identifying its rhetorical elements (the components of rhetorical knowledge).

Key Components of Rhetoric
- The **author** is the individual or group writing the piece.
    - Authors who are expert in a topic and who are writing for knowledgeable audiences may use **jargon**, which is vocabulary specific to a particular field.
    - An author may use more Extreme keywords if he is passionate about the topic at hand. An author may use more Moderating keywords if he is less invested or knowledgeable.
    - An author may modify his voice if writing on behalf of a group.
- The **audience** refers to the person or persons the author intended to read or hear the work.
- The **message** is the actual text written by the author, including factual information and explicit opinions.
    - The author's **voice** refers to how she uniquely selects words to deliver a message. It is how an author expresses her thoughts and can be unique and identifiable.
    - The **genre** is the category which the written work belongs to, such as a book, article, essay, letter, and so on.
    - The **medium** is the delivery system into which the written work can be placed, such as print, broadcast, and digital media.
- The **goal** is the reason why the author wrote the work.
    - The goal of many passages on the MCAT is to be persuasive, that is, to convince the reader to adopt new beliefs or to take action.
    - Other passages may have a goal of evoking an emotional response.
    - The author's **tone** is indicative of the goal of a written work because it reflects the author's attitude toward the subject matter.
    - Authors may write with more than one goal in mind.

- **Context** refers to two different (but related) concepts within rhetoric.
  - Within a written work, context refers to surrounding material that can be used to figure out the definition or significance of a particular element in the work. For example, a word's definition may be inferred from the other words in the same sentence; a paragraph's importance can be determined by comparison with nearby paragraphs.
  - Context can also refer to the greater social, political, economic, cultural, and intellectual environment in which the work was written.

Rhetorical Analysis

- Rhetoric depends on the communication skills of both the author of the message and the audience receiving the message.
- The **Aristotelian triad** describes three strategies, or appeals, used by an author to effectively transmit his message.
  - *Logos*, or logical persuasion, appeals to the audience's rational judgment.
  - *Ethos*, or ethical persuasion, appeals via the author's credibility.
  - *Pathos*, or emotional persuasion, appeals to the audience's feelings.

3

Keywords

3: Keywords

In This Chapter

3.1 Reading Strategically with Keywords	**28**	**3.3 Author Keywords**	**34**
Read for Content	28	Positive *vs.* Negative	34
Read for Organization	29	Extreme	35
Read for Perspective	29	Moderating	35
Read for Reasoning	30	Accounting for Opposition	36
3.2 Relation Keywords	**30**	**3.4 Logic Keywords**	**37**
Similarity	30	Evidence and Conclusion	37
Difference	31	Refutation	38
More Complex Relationships	32	**Concept and Strategy Summary**	**40**
		Worked Example	**42**

Introduction

> **LEARNING GOALS**
>
> After Chapter 3, you will be able to:
>
> - Apply keyword strategy within a passage to identify passage organization, perspective, and reasoning elements
> - Identify the relationship of a sentence to the text as a whole using Relation keywords
> - Use Author keywords to associate author tone and opinion with text
> - Connect evidence and conclusion within a passage by identifying Logic keywords

One of the biggest mistakes you can make as a student is to think that learning how to read is a one-time occurrence, like a switch that, once flipped, fully illuminates the darkened recesses of illiteracy and ignorance. In truth, it is always possible to improve your ability to read, both by refining your current approach and by broadening your capacity to read a range of texts in a variety of settings. Would you read a novel for pleasure the same way you read a textbook for homework? Once you recognize that there are many ways to read and multiple levels on which the written word can be appreciated, then you can learn to customize your reading approach to fit your particular purpose, whether it be relaxing with a piece of fiction or reaching for that higher score on the MCAT!

MCAT Critical Analysis and Reasoning Skills

> **Key Concept**
>
> Keywords are words and phrases commonly employed in CARS passages that serve as valuable clues for answering the accompanying questions. They fall into three broad categories: Relation, Author, and Logic.

In this chapter, you'll learn new ways to read strategically by paying attention to special terms that we call **keywords**. First, Relation keywords highlight the connections between ideas in passages: similarities, differences, and other relationships. Next, Author keywords are vital for providing insight into the mind of the passage's author. Finally, Logic keywords elucidate the reasoning in a text, which is the most frequently tested aspect of passages in the *Critical Analysis and Reasoning Skills* (CARS) section. We begin with a discussion of how to read using these vital clues before examining each of the three major categories in turn. In the chapter that follows, the discussion of keywords will continue as their place in the Kaplan Method for CARS Passages is explained and demonstrated through the analysis of an example CARS passage, so treat Chapter 4 of *MCAT CARS Review* as a companion to this one.

3.1 Reading Strategically with Keywords

When it comes to reading dense academic prose—and this describes just about every passage in the CARS section—there are at least four ways to approach the text; that is, there are four distinct levels on which it might be appreciated. The first of these modes, **content**, is what you're most likely accustomed to looking for, whether reading for work or pleasure. On Test Day, you will want to pay attention to informational content, but you'll also want to broaden your focus to encompass the other three modes of reading: **organization**, **perspective**, and **reasoning**.

> **Key Concept**
>
> Any CARS passage can be understood in four different ways, which we call the modes of reading. Each mode answers at least one vital question:
> - Content—*What does the text say?*
> - Organization—*How do sentences connect? How do ideas relate?*
> - Perspective—*Why does the author write? How does the author feel? Who else has a voice?*
> - Reasoning—*How are claims supported? How are claims challenged?*

READ FOR CONTENT

Reading for content simply involves extracting the information from the text, discovering precisely *what* is being said. This is akin to looking for the message of the passage, as described in Chapter 2 of *MCAT CARS Review*. It is important to note that you will never be expected to have preexisting familiarity with the content of a passage in the CARS section; all information necessary to answer the questions is contained within the passage itself. This situation is in stark contrast to passages in the three science sections, in which you are expected to integrate outside knowledge with new information provided in the passage.

When reading for content, ask yourself questions like: *What is the author saying? What is the main idea of this passage? What topics are being explored? What opinions has the author stated in the passage?* We will not discuss specific lists of keywords when reading for content; instead, we will focus on buzzwords, as we do in the sciences. **Buzzwords** include proper nouns, names, dates, new terms, and jargon.

READ FOR ORGANIZATION

Although there are many ways to talk about the "organization" of a passage, we are specifically referring to the ways in which the different ideas presented in the passage relate to one another. If the informational content is the *what* of the text, then the organization is the *how*. As you'll see in the next chapter, this organizational structure will be reflected in the Outline you construct on your noteboard, as you Label each paragraph to account for its function within the larger passage.

As you read for organization, you'll be considering questions like the following: *How does what I'm reading now connect to what came before? How many distinct ideas are actually being presented in this paragraph? Is this a new concept or just a restatement of a previous one? Where is the author going with this? How does this new point connect to the author's thesis?* Answering these questions becomes much more manageable when authors use Relation keywords. You may have learned about these in high school English or first year composition courses as "transition words" because they facilitate the movement (or transition) between sentences.

READ FOR PERSPECTIVE

In addition to watching out for the organization of ideas, you'll also want to pay attention to the different perspectives contained in the passage. Many authors of the passages used in CARS attempt to conceal their biases and mask their opinions as facts; they rarely state their intentions overtly. Therefore, it is essential to attend to the rhetorical aspects of the text presented in Chapter 2 of *MCAT CARS Review*—especially the goal, tone, and voice. The key to reading for perspective is thinking about the *why* behind the *how*. Consider the author's Goal in writing the passage and how each part of the passage functions in achieving that larger objective.

While every text embodies its author's perspective, many passages contain alternative voices as well, often noted by references to particular thinkers like *Carol Gilligan* or *Ludwig Wittgenstein* or to schools of thought like *postmodernism* or *American pragmatism*. On other occasions, divergent perspectives are indicated more indirectly by reference to a *traditional view* or *critics* or, even more subtly, with language like *some contend* or *others have argued*. In passages that feature multiple viewpoints, you are virtually guaranteed to see questions that require you to sort out exactly *who* said *what*.

Reading for perspective means thinking about the following: *Why is the author writing this? What is the purpose of this part of the passage? How many distinct voices are there in the passage? Does the author agree with what is said in this sentence, or does this sentence represent some other voice? How does the author really feel about this?* The second type of keywords, called Author keywords, thrust you inside the mind of the writer, letting you see things from her perspective.

MCAT Critical Analysis and Reasoning Skills

READ FOR REASONING

Finally, the fourth way to read a passage is to examine the structure of its reasoning. In just about every passage that you'll encounter on Test Day, the author will attempt to persuade the audience to adopt certain beliefs. To do so, it's not enough simply to repeat the claim that she wants you to accept. Rather, the author provides supporting reasons, or additional statements that make those beliefs more plausible. Giving reasons is also known as arguing, argumentation, or making an argument—what we called *logos*, or logical appeals, in Chapter 2 of *MCAT CARS Review*. You'll find that the logical structure of a text might be quite different from its organization; for instance, the so-called "conclusion" of the author's main argument could come in the first sentence of the passage.

While reading for reasoning, ask yourself questions like: *What is the author trying to convince the audience to believe? Does the author give any reasons to believe this sentence is true? Does this new claim make a previous one more plausible, or does it perhaps challenge an earlier one? What evidence or refutations does the author provide for this opinion? Which claims in the passage are given the most support?* Logic keywords, the final category, will aid you in sorting out the structure of the argument. Furthermore, argumentation and logic are such crucial topics for Test Day that they constitute the focus of both Chapters 5 and 6 of *MCAT CARS Review*—a necessary supplement to the discussion of Logic keywords later in this chapter.

3.2 Relation Keywords

When tackling a CARS passage, it is essential to be able to understand how what you're reading now fits into the whole—what we just called reading for organization. While there are many ways in which ideas might be related to one another, the vast majority of **Relation keywords** will fall into one of two subcategories: Similarity or Difference.

SIMILARITY

Keywords that indicate relations of similarity include *and*, *also*, *moreover*, *furthermore*, and so on. In addition, phrases that indicate examples (*such as*, *for instance*, and *take the case of*) fall into this category, as do demonstrative pronouns (*this*, *that*, *these*, *those*). Generally speaking, when a sentence or clause begins with a **Similarity keyword**, it will be continuing in the same vein as what came before. In other words, it won't be saying anything particularly new. As a consequence, if you understood the preceding material, you can generally read a clause that follows a Similarity keyword quickly, briskly moving through the text until a new keyword signals something different. On the other hand, if you struggled to understand a

> **MCAT Expertise**
>
> If you understand what an author is saying and encounter a Similarity keyword, skim the following text. You know you'll be seeing more of the same idea, and you may save a few seconds of time—those seconds add up!

particular sentence but see a Similarity keyword at the beginning of the next sentence, keep moving forward—the author is likely to continue with the same idea but may rephrase it in easier words.

Even though they are not technically key*words*, certain punctuation marks can also indicate that a similar idea is coming up. Most notably, colons (:) and semicolons (;) are commonly employed to function as the verbal equivalent of the equals sign (=). The use of dashes—such as the ones surrounding this clause—and parentheses (the marks enclosing this phrase) also tend to indicate elaboration upon the same general theme. Finally, keep an eye out for quotation marks (" and "): while quotes can serve a variety of functions, one of the most common is to use another person's voice to restate the point the author just made.

DIFFERENCE

In contrast, **Difference keywords** will usually merit additional attention when you encounter them because they tend to suggest more interesting (and therefore testable) relationships than similarity or continuity. Common Difference keywords like *but*, *yet*, *however*, *although*, and *otherwise* signal a change in the direction of the text. Like Similarity keywords, Difference keywords serve as transitions between sentences, but they can also indicate deeper conceptual relations—the differences between two solutions to a political problem, a point of disagreement between various critics of a literary work, or a rapid change in opinion from one time period to another, to give just a few examples. The connections between ideas are among the most commonly tested aspects of CARS passages, so strive to understand these relationships with as much specificity as the given clues allow.

While punctuation symbols often indicate similarity, this is not always true. If a punctuation symbol is accompanied by another type of keyword, the actual word or phrase generally takes precedence in determining the relationship (for example, a semicolon followed by *yet* usually suggests a point of difference rather than similarity). This points out the greater importance of anticipating while you read; a great score in CARS is predicated on your ability to be a critical reader rather than a passive reader. When we passively read for pleasure, we tend to glide over text and often only understand the superficial message of the written work. In critical reading, we are continuously questioning the text and setting expectations for where the author will go with an idea. Even if they are not met, it is still worthwhile to set these expectations. In fact, when an author takes a starkly different route in a part of the passage than we expect, the testmakers are more likely to ask about this text. That which is rhetorically unusual in the passage and frustrates expectations becomes excellent material for MCAT questions.

MCAT Expertise

When you encounter a Difference keyword, slow down your reading slightly. These keywords signify a change in the author's focus or a direct contrast between two things; either way, we need to know how the trajectory of the passage is changing to keep a step ahead of the author.

MORE COMPLEX RELATIONSHIPS

While the keywords considered under this heading might broadly count as Difference keywords, they designate special types of divergence and are especially ripe for CARS questions. While by no means exhaustive of the types of relations you might see on Test Day, Oppositions, Sequences, and Comparisons are three of the most common.

Oppositions

Words like *not*, *never*, *on the contrary*, and *as opposed to* indicate not merely a difference but an outright **Opposition** or conflict between ideas. Many authors of passages used in CARS like to create **dichotomies**, which are divisions of entities into two categories. These categories are considered **mutually exclusive**, meaning that they don't overlap. The use of *either...or*, *on one hand...on the other hand*, and similar parallel phrase constructions are good indications of a **dualism**. Often, but not always, these dichotomies will be depicted as **exhaustive**, meaning that everything falls into one of the two categories. For example, an author writing about human behavior might claim that *all actions are either free choices or involuntary reflexes*, leaving no space for shades of gray.

Sequences

Some Relation keywords suggest a series of events advancing in time: *initially*, *first*, *second*, *third*, *next*, *subsequently*, *before*, *after*, *last*, *finally*. These words will usually be spaced relatively evenly throughout the passage or at least throughout a paragraph or two, so note how they organize the text into chunks. **Sequences** are something of a hybrid between Similarity and Difference, with each word suggesting not only a connection to a larger process but also a departure from the other steps in the series. Generally, you're better off taking your time with these, at least until you have a good idea of how the sequence will unfold.

Some sequences can set up a clear difference between time periods. *Historically*, *traditionally*, *used to*, *originally*, and—when used in comparison to a later time—*initially* and *before* can be used as time-based Difference keywords when contrasted with words like *now*, *currently*, *modern*, *later*, and *after*. Such a setup often implies that new information was learned or discovered in the intervening time: *The traditional interpretation vs. A more modern understanding* or *Historically, we thought vs. But now, we know*.

Comparisons

Sometimes authors will evaluate ideas and rank them relative to other ideas. More often than not, authors will consider only two concepts at a time, contrasting them through the use of **Comparison keywords** like *more*, *less*, *better*,

---

Bridge

Mutual exclusivity is also an important condition in some probability problems. If two potential outcomes are mutually exclusive, the probability of both occurring, $P(A \text{ and } B) = 0$. If they are also exhaustive, then the probability of either occurring, $P(A \text{ or } B) = 1$. Probability rules are discussed in Chapter 12 of *MCAT Physics and Math Review*.

and *worse*. That said, occasionally authors will compare three or more items, or offer vague judgments of superiority (or inferiority) of one item over all others, reflected by superlatives such as *most*, *least*, *best*, and *worst*. When revealing attitudes, Comparison keywords function more like Author keywords, further explained below.

Table 3.1 lists examples of Relation keywords in each category. Note that some words can fit into more than one category; for example, *not* reveals a difference, but it can also indicate a direct opposition.

MCAT Expertise

Simply memorizing lists of keywords is not sufficient to extract all the information from a passage. A Difference keyword can serve slightly different functions depending on the context in which the word is found. An Evidence keyword may actually provide very little support for the logic of a passage, depending on how the author structures her argument. Recognize that MCAT CARS success is not about memorizing the fact that *but*, *yet*, and *however* are Difference keywords but is instead about understanding how Difference keywords affect our interpretation of the author's message.

Similarity	Difference	Opposition
and	but	not/never/none
also	yet	either … or
moreover	however	as opposed to
furthermore	although	on the contrary
like	(even) though	versus (*vs.*)
same/similar	rather (than)	on one hand … on the other hand
that is	in contrast	otherwise
in other words	on the other hand	**Sequence**
for example	otherwise	before/after
take the case of	nevertheless	earlier/later
for instance	whereas	previous/next
including	while	initially/subsequently/finally
such as	different	first/second/third/last
in addition	unlike	historically/traditionally/used to
plus	notwithstanding	now/currently/modern
at the same time	another	**Comparison**
as well as	instead	better/best
equally	still	worse/worst
this/that/these/those	despite	less/least
: [colon]	alternatively	more/most
; [semicolon]	unless	–*er*/–*est*
— [dash]	not	primarily
( ) [parentheses]	conversely	especially
" " [quotes]	contrarily	above all

Table 3.1. Common Relation Keywords

3.3 Author Keywords

Author keywords can be among the most subtle clues that you'll encounter on Test Day, but they are crucial for answering the many questions you'll face that ask about the author's attitudes. Authors of passages used in CARS rarely say *I believe* or *it seems to me* (and if you do find this language, it's more often in the humanities than the social sciences). Instead, they are more likely to hint at their opinions by selecting verbs, nouns, adjectives, and adverbs that carry a particular emotional valence—a connotation of either approval or disapproval. Moreover, authors will use characteristic words and short phrases to make their claims more Extreme (indicating emphasis and strengthening ideas), as well as others that Moderate their claims (qualifying or limiting what they are saying and weakening ideas).

POSITIVE *vs.* NEGATIVE

Understanding the author's attitude becomes a much simpler matter if we employ a metaphor taken from the sciences. Just as an atom or molecule might possess an electrostatic charge, so too can a word or phrase contain a kind of emotional charge that may be positive or negative. Because most of the language in CARS passages tends to be "uncharged" or neutral, you'll predominantly pay attention to the exceptions—those cases in which terms have clear positive or negative connotations.

Positive keywords include nouns such as *masterpiece*, *genius*, and *triumph*; verbs such as *excel*, *succeed*, and *know*; adjectives such as *compelling*, *impressive*, and *elegant*; and adverbs such as *correctly*, *reasonably*, and *fortunately*. Among **Negative keywords** would be nouns such as *disaster*, *farce*, and *limitation*; verbs such as *miss*, *fail*, and *confuse*; adjectives such as *problematic*, *so-called*, and *deceptive*; and adverbs such as *questionably*, *merely*, and *purportedly*.

Keep in mind that, just as there is a difference between a cation with a +1 charge and one with a +3 charge, so too is there a difference between a moderately positive opinion and an extremely positive one. For instance, an author probably approves more strongly of a novelist described as *a masterful artist* than one portrayed merely as *a quality writer*. Consequently, it may be helpful to think of the author's attitude as varying along a spectrum or continuum, with extremely positive opinions on the one end and extremely negative on the other, as in Figure 3.1. Note that most authors' attitudes in CARS fall in a comfortable middle ground between being too extreme and being too moderate, as implied by the relative widths of the sections in the diagram. When Outlining a passage, as described in Chapter 4 of *MCAT CARS Review*, the author's attitude can be written in shorthand: + for moderately positive, − − − for extremely negative, and so on.

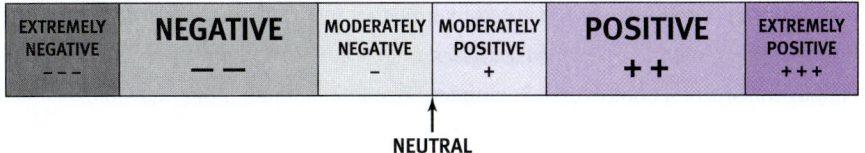

Figure 3.1. The Spectrum of Author Attitudes (Linear)

> **MCAT Expertise**
> Most passages on the MCAT contain strong, but not extreme, opinions. Rarely will an author be completely neutral because there is little reasoning for the questions to test you on if the author does not express at least a moderately positive or negative opinion.

Note that in addition to positive, negative, or neutral, an author can also be **ambivalent**. Ambivalence literally means *feeling both ways*, and it is as different from **impartiality**—having no strong opinion one way or another—as the set of 1 and −1 is from the number 0. Continuing the analogy with electrostatic charge, an ambivalent attitude is like an amino acid in its *zwitterionic* form, with both a positively charged and a negatively charged end, and an impartial attitude is like an uncharged, unpolarized atom. Describing the net zero charge of a zwitterion as merely "neutral" would mean neglecting its distinctive properties. The MCAT won't let you get away with such oversimplifications!

> **Key Concept**
> While they are both neutral overall, these two attitudes are very different:
> - Ambivalent = having both a positive and negative opinion
> - Impartial = having neither a positive nor negative opinion

EXTREME

Placing a particular idea on the author-attitude spectrum above becomes easier by paying attention to **Extreme keywords**, a type of Author keyword that you can imagine as enhancing the charge of what the author is saying, forcing the author into one or the other extreme. These words and short phrases are functionally equivalent to exclamation points (!), offering insight into what the author feels passionately about and regards as important.

Examples of Extreme keywords include *indeed*, *very*, *really*, *quite*, *primarily*, *especially*, *obviously*, *foremost*, *always*, *in fact*, *above all*, and *it is clear that*. Note that words that indicate necessity, like *need* and *must,* also serve as Extreme keywords, as do words that indicate value judgments like *should* and *ought*—these tend to be rare in CARS passages, so they deserve special consideration when they do appear.

MODERATING

Authors will sometimes modify the strength of their claims in the other direction by using qualifying language, also known as hedging. **Moderating keywords** are those words that set limits on claims in order to make them easier to support because a stronger statement is always more difficult to prove than a weaker one. For example, it would be an extreme claim to say that *human beings are motivated only by greed*. Though some might agree with this formulation, the bulk of MCAT authors would sooner water it down by saying something like *in many aspects of life,* *humans are predominantly motivated by greed*, or even further limit it to a subset of human beings, such as *investment bankers are often motivated by greed*. Such modifications transform a controversial claim into one that is much more plausible.

Among the most important Moderating keywords are those that use the language of possibility, such as *can*, *could*, *may*, and *might*. Claims about what is possible are always weaker than claims about what is definitely true. Other Moderating keywords include limits on time or place, whether stated specifically or in vague phrases such as *now*, *here*, *at times*, *in some cases*, or *in this instance*. Still others will impose general constraints on meaning; examples include *in this sense*, *according to this interpretation*, or *in a manner of speaking*.

ACCOUNTING FOR OPPOSITION

One final consideration when working with Author keywords is contradiction or opposition, a special type of Difference keyword mentioned above. It can be particularly tricky to figure out authors' attitudes in CARS when they use double negatives (or worse!). The key is to remember that the opposite of an extreme statement will typically be a moderate statement of the reverse charge. In the diagram below, which is just a slight rearrangement of the author attitude spectrum presented in Figure 3.1, a word of opposition will typically serve to flip the author's view 180 degrees.

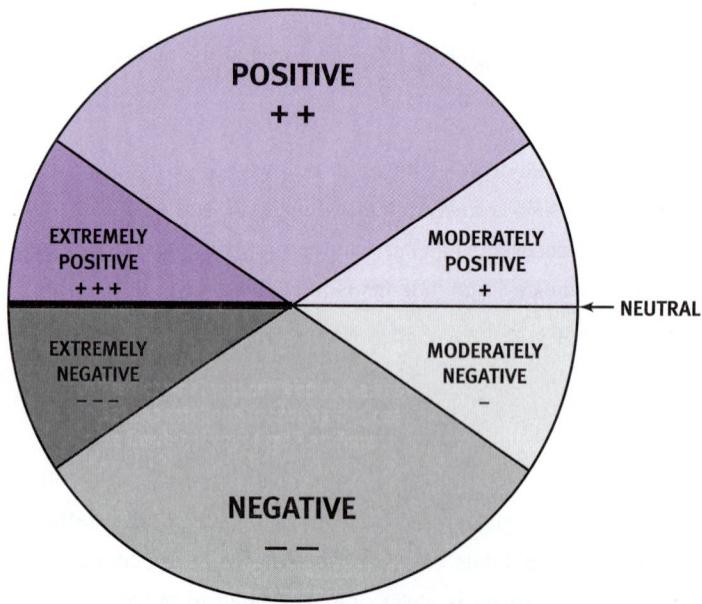

Figure 3.2. The Spectrum of Author Attitudes (Circular)

So, if an author were to claim that a particular event is *impossible*, this statement would fall somewhere in the extremely negative range. On the other hand, suggesting the situation was *not impossible* would be making a moderately positive claim, effectively saying that it was *possible*. Be aware, however, that some words retain their strength even when accompanied with *not* or some other word of opposition—for instance, *must* is extremely positive in charge, while *must not* is extremely negative in charge.

Table 3.2 lists examples of Author keywords in each category.

Positive	Negative	Extreme	Moderating
masterpiece	disaster	must	can/could
genius	farce	need/necessary	may/might
triumph	limitation	always	possibly
excel	miss	every	probably
succeed	fail	any	sometimes
know	confuse	only	on occasion
compelling	problematic	should/ought	often
impressive	so-called	indeed	tends to
elegant	deceptive	very	here
correctly	questionably	especially	now
reasonably	merely	obviously	in this case
fortunately	purportedly	above all	in some sense

Table 3.2. Common Author Keywords

3.4 Logic Keywords

The final level of reading (for reasoning) is perhaps the most difficult because the special, one-way relationship between a conclusion and its evidence is among the most complex you'll encounter on Test Day. Making matters worse is that **Logic keywords** tend to be relatively rare, occurring less frequently than either Relation or Author keywords in most passages. As a final complication, when they do occur, they are sometimes used to refer to causal connections rather than logical justifications. Notwithstanding these difficulties, Logic keywords are a powerful tool: once you gain proficiency in recognizing them and understanding what they entail, you'll find the many CARS questions on reasoning much less daunting.

EVIDENCE AND CONCLUSION

The precise nature of the relationship between the different parts of an argument will be considered in Chapter 6 of *MCAT CARS Review*, but for now you can use the following simple distinction: a **conclusion** is a claim that the author (or whomever the author is speaking for) is trying to convince the audience to believe, while pieces of **evidence** are the reasons that are given for believing it. Be sure to determine whenever you see the following keywords whether the conclusion is one that the author would endorse or whether it is intended to represent some other viewpoint by paying attention to nearby Author keywords and other clues.

Typical examples of **Evidence keywords** are *because*, *since*, *if*, *why*, *the reason is*, *for example*, *on account of*, *due to*, *as a result of*, *is justified by*, and *after all*. There is a bit less variety in **Conclusion keywords**, which also tend to occur less frequently than Evidence keywords. The most important conclusion words to know are *therefore*, *thus*, *then*, *so*, *consequently*, *leading to*, *resulting in*, *argue*, and *conclude*.

In Chapter 6 of *MCAT CARS Review*, the exact nature of logical support, also known as **justification**, will be discussed in depth. However, it's important to note that Evidence and Conclusion keywords sometimes signal a different relationship: cause and effect. Unlike a mere correlation, in which two occurrences are found to accompany one another, **causation** is a one-way relationship in which the first event (cause) always precedes the second (effect). Moreover, in contrast to sequences of events, which may happen coincidentally, causes are directly responsible for leading to their effects. The upshot of this is that whenever you encounter Evidence or Conclusion keywords, you should always clarify whether you are dealing with events in the world that have causal connections or with statements about the world that have logical ones.

REFUTATION

Refutation keywords will not always be included in the presentation of an argument, but they are effectively the opposite of evidence—countervailing reasons for rejecting a conclusion. Although rarer than other Logic keywords, Refutation keywords are important when they do occur for representing one of the most commonly tested aspects of logical reasoning. They include words such as *despite*, *notwithstanding*, *challenge*, *object*, *counter*, *critique*, *conflict*, and *problem*.

MCAT Expertise

Having trouble figuring out what part of the argument a Logic keyword is indicating? Try this simple substitution test: if you can replace the word or phrase with *because* or *because of*, then whatever follows is a piece of evidence. If, instead, *therefore* would preserve the meaning, the subsequent claim is a conclusion.

Table 3.3 lists examples of Logic keywords in each category.

Evidence	Conclusion	Refutation
because (of)	therefore	despite
since	thus	notwithstanding
if	then	challenge
for example	so	undermined by
why	consequently	object
the reason is	leading to	counter(argument)
as a result of	resulting in	critique/criticize
due to	argue	conflict
as evident in	conclude	doubt
justified by	imply	problem
assuming	infer	weakness
after all	suggest	called into question by

Table 3.3. Common Logic Keywords

Conclusion

By watching out for Relation, Author, and Logic keywords, you'll find yourself appreciating the text on multiple levels, going beyond the content to look at organization, perspective, and reasoning. In the subsequent chapter, we'll expand on the ideas introduced here, discussing in depth how you can use the insights gained from strategic reading to construct a scratch paper Outline of the passage, Labeling what is vital in each paragraph to make answering the questions more efficient.

MCAT Critical Analysis and Reasoning Skills

CONCEPT AND STRATEGY SUMMARY

Reading Strategically with Keywords

- **Keywords** help us figure out how information is related within a passage, or they may modify our interpretation of the text around them.
- When we read for **content**, we ask *What does the text say?*
  - Reading for content is guided by **buzzwords**, like proper nouns, names, dates, new terms, and jargon.
  - You are never expected to have prior familiarity with the content of a passage in CARS.
- When we read for **organization**, we ask *How do sentences connect?* and *How do ideas relate?*
  - Reading for organization is guided by **Relation keywords**.
  - The organizational structure of a passage should be Outlined on the available noteboard.
- When we read for **perspective**, we ask *Why does the author write? How does the author feel?* and *Who else has a voice?*
  - Reading for perspective is guided by **Author keywords**.
  - Other individuals' voices can be referred to specifically (by name, by category) or vaguely (*some* or *others*).
- When we read for reasoning, we ask *How are claims supported?* and *How are claims challenged?*
  - Reading for reasoning is guided by **Logic keywords**.
  - The logical structure of a passage may be quite different from its organization.

Relation Keywords

- **Relation keywords** show how what you're reading now fits into the passage as a whole.
- **Similarity keywords** indicate that the following material continues in the same vein as the preceding material.
- **Difference keywords** signal a change in the trajectory of the passage.
- **Opposition keywords** are particularly strong Difference keywords that create a dichotomy, or divisions of entities, into two categories.
  - **Mutually exclusive** categories do not overlap.
  - When dichotomies are **exhaustive**, all relevant entities fit into one or the other category.

- **Sequence keywords** suggest a series of events advancing in time. They may also be used to set up a contrast between two time periods.
- **Comparison keywords** rank ideas relative to each other.

Author Keywords

- **Author keywords** indicate the author's thoughts or opinions about the topic.
- **Positive** and **Negative keywords** indicate whether an author likes, agrees with, or supports the topic; dislikes, disagrees with, or opposes the topic. Authors can also be neutral.
  - An author with an **ambivalent** attitude has both positive and negative opinions on a topic.
  - An author with an **impartial** attitude has neither positive nor negative opinions on a topic.
- **Extreme keywords** enhance the charge of what the author is saying.
- **Moderating keywords** permit the author to qualify a claim, or hedge.
- The opposite of an extreme statement tends to be a moderate statement with the opposite charge.

Logic Keywords

- **Logic keywords** indicate the relationships between different parts of a logical argument.
- **Conclusion keywords** signal what the author is trying to convince the audience to believe.
- **Evidence keywords** describe the reasons why the audience should believe the author's claim.
- **Refutation keywords** provide reasons for rejecting a conclusion.

MCAT Critical Analysis and Reasoning Skills

WORKED EXAMPLE

Use the Worked Example below, in tandem with the subsequent practice passages, to internalize and apply the strategies described in this chapter. The Worked Example matches the specifications and style of a typical MCAT CARS passage.

If you haven't had a chance to read further in this book, you may not have read about the Kaplan Method for CARS Passages, called Outlining. You'll learn all about Outlining in Chapter 4 of *MCAT CARS Review*. Outlines help us remember what the author discussed in each paragraph, and the Goal statement reminds us of the author's overall purpose in writing the passage. A sample Label is provided for each paragraph.

Passage	Analysis
No one is eager to touch off the kind of hysteria that preceded the government's decision to move against Alar, the growth regulator once used by apple growers. When celebrities like Meryl Streep spoke out against Alar and the press fanned public fears, some schools and parents rushed to pluck apples out of the mouths of children. Yet all this happened before scientists had reached any consensus about Alar's dangers.	This passage starts out with some dramatic language: *hysteria*, *fanned public fears*, and parents who *rushed to pluck apples out of the mouths of children*. A stance isn't initially taken on the Alar incident, so we're on the lookout for an indication of the author's opinion. The last sentence of this paragraph provides a clue: it starts with a Difference keyword, *yet*, indicating that the author doesn't agree with public hysteria that might not be grounded in scientific evidence. A quick Label for this paragraph could be: **P1.** Alar and unwarranted public hysteria; Auth: reaction unfounded
Rhetoric about dioxin may push the same kind of emotional buttons. The chemical becomes relatively concentrated in fat-rich foods—including human breast milk. Scientists estimate that a substantial fraction of an individual's lifetime burden of dioxin—as much as 12 percent—is accumulated during the first year of life. Nonetheless, the benefits of breast-feeding infants, the EPA and most everyone else would agree, far outweigh the hazards. Now environmentalists say dioxin and scores of other chemicals pose a threat to human fertility—as scary an issue as any policy makers have faced.	More dramatic language shows that environmentalists are concerned that dioxin threatens fertility, *as scary an issue as any policy makers have faced*. Again, this paragraph uses a Difference keyword, *nonetheless*, to contrast the author's opinion with those of the frightened public: *nonetheless, the benefits of breast-feeding infants…far outweigh the hazards*. Here, our Label is: **P2.** Dioxin is similar case; Env: dioxin is a threat; Auth: reaction is emotional

Passage	Analysis
But in the absence of conclusive evidence, what are policy makers to do? What measure can they take to handle a problem whose magnitude is unknown? Predictably, attempts to whipsaw public opinion have already begun. Corporate lobbyists urge that action be put on hold until science resolves the unanswered questions. Environmentalists argue that evidence for harm is too strong to permit delay. This issue is especially tough because the chemicals under scrutiny are found almost everywhere.	This paragraph highlights the author's concern over the dilemma policy makers face when public opinion isn't supported by scientific evidence: *What measure can they take to handle a problem whose magnitude is unknown*? Two groups holding opposing points of view are trying to *whipsaw public opinion*; on one end, corporate lobbyists want more research to be done before policy decisions are made, and on the other, environmentalists urge immediate action. The challenge of solving this opposition is highlighted with the Extreme keywords *especially tough*. An appropriate Label might be: **P3.** Auth: likely no good response; Dioxin risk poorly understood
Because many of them contain chlorine or are by-products of processes involving chlorine compounds, the environmental group Greenpeace has demanded a ban on all industrial uses of chlorine. The proposal seems appealingly simple, but it would be economically wrenching for companies and consumers alike. With the escalating rhetoric, many professionals in the risk-assessment business are worried that once again emotion rather than common sense will drive the political process. "There is no free lunch," observes Tammy Tengs, a public health specialist at Duke University. "When someone spends money in one place, that money is not available to spend on other things." She and her colleagues have calculated that tuberculosis treatment can extend a person's life by a year for less than $10,000—surely a reasonable price tag. By contrast, extending a life by a year through asbestos removal costs nearly $2 million because relatively few people would die if the asbestos were left in place. That kind of benefit–risk analysis all too rarely informs the decisions made by government regulators.	The author is again critical of rash decisions involving chemical regulation. Greenpeace's argument is presented with the Evidence keyword *Because*; this group arrives at the conclusion to demand a ban on industrial uses of chlorine. The author derides this conclusion with Author and Refutation keywords such as *appealingly simple, but…economically wrenching for companies and consumers alike*. In the previous paragraph, the author wondered how tough decisions could be made. Here a solution is identified: *benefit–risk analysis*. Though the government is well intentioned in its efforts to regulate harmful chemicals, *benefit–risk analysis all too rarely informs [their] decisions*. All too rarely are Extreme Author keywords, indicating that this paragraph is a good one to consult for the author's opinion if needed later. A Label for this paragraph might be: **P4.** Common sense/risk analysis should guide decisions about threats

MCAT Critical Analysis and Reasoning Skills

Passage	Analysis
As the EPA raises anew the dangers of dioxin, the agency needs to communicate its findings to the public in a calm and clear fashion. John Graham, director of the Harvard Center for Risk Analysis, suggests that people should strive to keep the perils posed by dioxin in perspective and remember other threats that are more easily averted. "Phantom risks and real risks compete not only for our resources, but also for our attention," Graham observes. "It's a shame when a mother worries about toxic chemicals, and yet her kids are running around unvaccinated and without bicycle helmets."	The final paragraph uses Difference keywords such as *compete*, *not only… but also*, and *yet* to contrast overblown (or even *phantom*) dangers like dioxin with pressing *real risks* like unvaccinated children without bicycle helmets. The author would likely agree with John Graham, whose opinion is mentioned with the neutral Author keyword *suggests*. Graham argues that our resources and attention should be focused first on the health risks that we know exist. A Label here might be: **P5.** Perspective; focus on real risks, not imagined ones (John Graham)

Here's a sample Outline and Goal for this passage:

P1. Alar and unwarranted public hysteria; Auth: reaction unfounded

P2. Dioxin is similar case; Env: dioxin is a threat; Auth: reaction is emotional

P3. Auth: likely no good response; Dioxin risk poorly understood

P4. Common sense/risk analysis should guide decisions about threats

P5. Perspective; focus on real risks, not imagined ones (John Graham)

Goal: To discuss the alleged threat posed by dioxin and to argue in favor of risk assessments, not emotion, to guide policy decisions

3: Keywords

Question	Analysis
1. According to the passage, it is dangerous to react drastically to recently posed health hazards for all of the following reasons EXCEPT:	Scattered Detail questions like this one have three incorrect answers that appear as details in the passage plus the correct answer that isn't mentioned. The question can be systematically answered by evaluating each answer choice in turn until the correct answer—the one not mentioned in the passage—is found.
A. proven precautions are overlooked.	**(A)** is found in paragraph 5; it's more effective to worry about safety methods that have been proven to improve safety, like bicycle helmets and vaccines.
B. public fear leads to irrational action.	**(B)** is discussed throughout the passage. In paragraph 1, Alar is mentioned as a chemical that was attacked as a result of public outcry without the backing of scientific evidence.
C. insurance premiums will increase.	**(C)**, *insurance premiums*, is not mentioned in the passage, making it the correct answer. Once the correct answer is identified, there's no need to continue evaluating incorrect answers.
D. economic burdens can occur.	**(D)** is mentioned in paragraph 4.

MCAT Critical Analysis and Reasoning Skills

Question	Analysis
2. In the context of the passage, the author uses the term "whipsaw public opinion" (paragraph 3) to refer to:	Helpfully, this Definition-in-Context question points to the third paragraph, so the answer will be found there. Who wants to *whipsaw public opinion* in paragraph 3? *In the absence of conclusive evidence*, corporate lobbyists and environmental groups are both trying to convince the public to accept an extreme view of the risks of dioxin. Whipsaw must mean something like *advocating for an extreme opinion*.
A. changing the needs of the community.	**(A)** is Out of Scope because the *needs of the community* are never mentioned—only opinions.
B. convincing citizens to accept a polarized viewpoint on health hazards.	**(B)** matches the prediction; corporate lobbyists and environmental groups are trying to convince the public to take on a particular point of view regarding health hazards.
C. offering a variety of alternatives for health hazards.	**(C)** is a Distortion; the extreme groups mentioned in the paragraph don't offer a *variety* of alternatives.
D. acting irrationally in response to government policy.	**(D)** is also a Distortion; although the author might believe that extreme groups are *acting irrationally*, this isn't related to the attempt to change public opinion.

3: Keywords

Question	Analysis
3. For which of the following reasons does the author cite the Alar incident in paragraph 1?	For this Function question, determine why the Alar incident is discussed in paragraph 1. It's an example of a time when emotional public opinion surpassed a lack of scientific evidence. It's a historical precedent for the current debates about dioxin and chlorine that are discussed later in the passage.
A. To show the bureaucracy involved in changing a chemical plant's mode of operation	(A) is Out of Scope; chemical plant operation is never mentioned.
B. To illustrate the problem in publicly announcing health hazards before conclusive scientific evidence has been formulated	(B) matches the prediction, highlighting the contrast between public opinion and lack of scientific evidence.
C. To show that drastic reaction is often the best way to solve a crisis	(C) is an Opposite; the author thinks that drastic reaction is an awful way to solve a crisis.
D. To demonstrate that it takes a celebrity to effect public change	(D) is a Distortion; celebrities like Meryl Streep helped build public opinion against Alar, but nowhere is it suggested that *it takes a celebrity to effect public change*, and the role of celebrities is otherwise not the focus of paragraph 1.

MCAT Critical Analysis and Reasoning Skills

Question	Analysis
4. Which of the following statements, if true, would most WEAKEN the author's argument?	We can quickly remind ourselves of the author's main argument by referring to the Goal statement in our Outline. The Goal for this passage is *To discuss the alleged threat posed by dioxin and to argue in favor of risk-assessments, not emotion, to guide policy decisions*. In other words, the author argues that health fears are often exaggerated and should be subject to careful study. We want to weaken the author's argument, so we're looking for an answer choice to this Strengthen–Weaken (Beyond the Passage) question that either says that health fears are *not* exaggerated, or that they should *not* be subject to careful study.
A. The EPA carefully considered the research results of a highly qualified team of scientists, economists, and public policy makers who researched the asbestos and Alar threats before any governmental action was performed.	**(A)** is correct; it counters the author's argument because it says that the EPA did in fact have substantial scientific evidence that asbestos and Alar were harmful before the agency took action.
B. Large numbers of babies have been born with defects over the last 20 years when levels of Alar have been extremely high.	**(B)** is not an aspect of the author's argument; the author never claims that Alar is completely harmless, only that further study needs to be done. If research showed that Alar does cause birth defects, then the author would accept those results without them weakening the argument.
C. Activist groups, such as Greenpeace, believe that the use of chemicals in our society has reached overwhelming proportions and needs to be regulated immediately.	**(C)** is a statement that the author would agree with, and the passage already describes Greenpeace's extreme views.
D. Corporate lobbyists consider economic factors that may make certain precautions economically unfeasible.	**(D)** is a statement that the author would agree with; paragraph 3 mentions that corporate lobbyists want to halt regulatory action, and paragraph 4 notes that banning industrial uses of chlorine would have significant economic effects.

3: Keywords

Question	Analysis
5. All of the following are mentioned by the author in the passage in support of the main argument EXCEPT:	This is a Scattered Function question, where the three incorrect answers will be found in the passage but the correct answer will not. The incorrect answer choices might not use the exact same language as found in the passage, but they will still indicate an idea that's clearly present in the text. We will answer this question by locating and eliminating the three incorrect answers.
A. the idea that people often overlook health threats for which we already possess remedies.	**(A)** is mentioned at the very end of the passage. The author quotes John Graham, who contrasts *phantom risks and real risks*, and points to bicycle helmets and vaccines as viable solutions to real health risks. Eliminate this answer choice.
B. biased groups will try to sway citizens into believing that their stance is the only correct way of handling health hazards.	**(B)** can be found in paragraph 3; corporate lobbyists and environmentalists are two polarized groups that are attempting to sway public opinion in the absence of concrete evidence, so eliminate this answer choice.
C. public reaction has led to unnecessary actions that have wasted time and money.	**(C)** is supported by the author's statements in paragraph 4 about the dubious merit of asbestos removal. Extending lives through asbestos removal is much more expensive (and implicitly time-consuming) compared to other endeavors like tuberculosis treatment.
D. chemicals in food and homes have caused too many deaths in modern society.	**(D)** is not mentioned by the author in the passage, so it is the correct answer. The author does argue that the harmful impact of chemicals has been overblown, but not that chemicals have caused many deaths.

Practice Questions

Passage 1 (Questions 1–5)

Since the transition of governmental structure from the Articles of Confederation to the Constitution in 1789, the president has been granted the executive authority to veto legislation passed by Congress. The threat of a veto in many cases precipitates compromise on the content of a bill that would be otherwise mired in debate before it reached the president. The "regular" veto is a qualified negative veto, which necessitates a two-thirds vote by Congress in order to be overridden. The "pocket" veto, on the other hand, is exercised when a bill sits on the president's desk without being signed before Congress has adjourned (and is therefore unable to override the veto). Opponents of the pocket veto allege that its absolute nature grants the president excessive power. They liken it to a prerogative of the English kings that the Framers vehemently despised. The argument also embraces a vast body of commentary on the "Imperial Presidency"; that is, the growing accumulation of power in the executive relative to the legislative branch.

These arguments, in claiming an imbalance of federal powers through the subservience of the legislative bodies to the whims of the executive branch, misrepresent the pocket veto. Unlike the royal prerogative, against which the creators of this great nation so nobly rebelled, the pocket veto is exercised by a democratically elected leader pursuant to a clearly defined constitutional procedure in which presentation of a bill by Congress may be arranged so as to thwart the possible execution of the pocket veto. Moreover, an absolute veto forecloses further action on a proposal, whereas Congress may overcome a pocket veto by instituting a reintroduction and passage of the rejected bill in a subsequent term.

The "Imperial Presidency" developed from the encroachment of executive action into areas where it has been assumed that the legislative branch retains supremacy. The legislative process, however, clearly orders shared responsibility between the president and Congress. One should not mistake presidential powers granted to block legislation for those that would, in effect, supplant congressional authorization. The latter threatens to override the constitutional system of checks and balances; the former situation, typified by the pocket veto, is a part of that system of checks and balances—not surprisingly, given the original purpose and mandates of this historic document.

The arguments raised in *Kennedy v. Sampson* and *Barnes v. Kline* implicitly claim that a regular veto would be overridden, or not exercised at all. Consequently, the pocket veto grants the president a special political tool against "popular will" as exercised by Congress. Herein lies the fundamental disagreement over the pocket veto. Opponents press for the president to defer to a seemingly inevitable congressional victory, while proponents of this second type of veto stand behind its historical use by the president to stall or delay legislation he thinks unwise. If circumspection and deliberation are the more valued aspects of the lawmaking process, even the most blatantly political use of the pocket veto passes muster. Historical practice favors the president's role as an interloper.

1. The passage suggests that which of the following would be the likely consequence if overwhelmingly popular legislation is deferred by the president's use of the pocket veto?

   A. The vetoed legislation would be reintroduced by Congress.
   B. Congress would be powerless to pass similar legislation.
   C. The president would override the system of checks and balances.
   D. The pocket veto would be detrimental to future legislative efforts.

2. The author suggests that opponents of the pocket veto would most likely agree that:

   A. the president should not be allowed to exercise legislative authority.
   B. use of the pocket veto unfairly removes power from the legislative branch.
   C. Congress should have the right to override the pocket veto.
   D. the absolute veto should be reinstated by Congress.

3. The author would consider a "blatantly political use of the pocket veto" (paragraph 4) to be:

   A. unjustified because the will of the congressional majority should be respected.
   B. unwise because the president should be perceived to stand above partisan politics.
   C. appropriate if the president has pledged in advance to block the legislation in question.
   D. legitimate because it can force further consideration of a bill the president opposes.

4. It can be inferred that the author considers which of the following to be the strongest argument against the positions opposing the pocket veto?

   A. A regular veto of the legislation in question in *Kennedy* and *Barnes* would not have been overridden.
   B. A regular veto of the legislation in question in *Kennedy* and *Barnes* would probably have been overridden.
   C. The president would have been unlikely to use a regular veto because of fear of public opinion.
   D. In certain cases, the Constitution allows the president to delay legislation, which has majority support.

5. The author would be most likely to agree with which of the following statements about the "Imperial Presidency"?

   A. It represents an unprecedented threat to the continuity of American institutions.
   B. It is more in keeping with the present English system of government than with the American.
   C. The pocket veto is not really an example of tendencies toward an "Imperial Presidency."
   D. It has been the cause of increasingly frequent use of the pocket veto.

Passage 2 (Questions 6–10)

Five times as many workers may be needed to construct a power plant as to operate it. The numbers may be even more disproportionate for a major pipeline or dam. When the construction ends, a substantial reduction in population is virtually guaranteed. Hence, there may be no justification for providing an infrastructure necessary to maintain adequate levels of service during the construction period.

Money necessary to build water systems, schools, and roads, and to fund salaries and maintenance costs, is mismatched by traditional taxing programs. The construction project is usually not subject to local property tax until it nears completion, which may be five years after the impact has occurred. Alternative sources of tax revenue cannot begin to cover the cost of providing the necessary services. Even if some governments have money, they may not be the right governments. Some entities may suffer the impact of development without being able to tax it. For example, a development may be located in the county just outside the limits of an incorporated city. The county will be entitled to tax the property while the city may receive most of the project population and demand for services.

The 1960s and 1970s witnessed a new boomtown era in the West. The typical contemporary boomtown is fueled by a quest for energy in the form of a fossil-fueled electric generating plant, a hydroelectric dam, or a new mine. The energy project is typically located near a small community or is forced to start a community from scratch. Often, the boomtown is poorly planned and under-financed. Longtime residents find their community changed for the worse and newcomers find the town an undesirable place to live.

The boomtown is characterized by inadequate public services, undesirable labor conditions, confusion in community structure, and deterioration of the quality of life arising from rapid population growth due to a major economic stimulus. Accelerated growth is the most distinguishing characteristic of a boomtown.

Studies have shown that large-scale development in sparsely populated areas causes major social problems. Housing, street and water system construction, school development, and police and fire protection lag far behind population growth. Rent and property tax increases join with a rise in the general cost of living to harm persons on fixed incomes. Education in the community may suffer. One result of boomtown living is higher incidence of divorce, depression, alcoholism, and attempted suicide. Until recently, planners have ignored or understated such problems. While the boomtown promotes an "us against them" mentality—the old timers *vs.* persons brought to the community by the boom—the fact remains that all parties suffer. Newcomers may blame old-timers for a lack of support just as old-timers may blame them for a deterioration of community life. Consequences of the boomtown also harm the project developer. The undesirable community results in poor worker productivity and frequent worker turnover, factors that delay construction and push projects over budget. Problems of rapid growth in some boomtowns are compounded by the fact that most of the population disappears with the completion of project construction.

6. The passage suggests that there is often a lack of services associated with boomtowns. The author claims that all of the following are possible causal factors for the lack of services associated with a boomtown EXCEPT:

   A. the expected loss of a substantial number of residents after the completion of a project.
   B. lack of support from long-time residents.
   C. the location of an energy project just outside the limits of an incorporated city.
   D. the time lag between the beginning of project construction and the onset of tax payments for it.

7. Based on information provided in the passage, improved public services in boomtowns could result from which of the following?

   A. Establishment of an adequate infrastructure during project construction
   B. Decreased county funding for construction projects
   C. Better enforcement of tax programs
   D. Limiting services to the anticipated levels necessary for towns' long-term needs

8. The tone of the author's discussion of traditional taxing programs in regard to boomtowns can best be described as:

   A. outraged.
   B. concerned.
   C. disbelieving.
   D. complacent.

9. The author would be most likely to agree with which of the following statements concerning community life in a boomtown?

   A. Old-timers suffer the most from the new developments that occur because of energy project construction.
   B. A smaller number of boomtown residents would suffer from depression or alcoholism if planners did not underestimate such problems.
   C. Project developers would experience less worker turnover if they acknowledged the complaints of long-time residents.
   D. An "us against them" mentality is unproductive because all residents suffer from a boomtown's failings.

10. Consider the fifth paragraph independent from the passage. Which of the following best describes the organization of that paragraph?

    A. A finding is cited and then discussed.
    B. A prediction is made but then qualified.
    C. A point of view is set forth and then justified.
    D. A proposal is presented and then dismissed.

Explanations to Practice Questions

Passage 1 (Questions 1–5)

Sample Passage Outline

P1. Regular *vs.* pocket veto; Opponents: too much power, "Imperial Presidency"

P2. Auth: Opponents wrong; pocket veto by elected leader can be overridden by later reintroduction

P3. Pocket veto is part of system of checks and balances

P4. Pocket veto is tool against popular will supported by historical use; opponents' and proponents' arguments

Goal: To discuss the pocket veto and arguments on whether it allows the President too much power

1. A

What does the author present as a response to a veto of popular legislation? In paragraph 2, the author says that the pocket veto isn't absolute because legislation can be reintroduced at a later time. Thus, **(A)** is a likely course of action. **(B)**, on the other hand, is an Opposite. The author argues that the pocket veto is acceptable precisely because Congress can pass similar legislation at a later time. Similarly, **(C)** is an Opposite because the author argues in paragraph 3 that the pocket veto is a valuable part of the system of checks and balances, not a threat to it. Finally, **(D)** is also an Opposite. The author argues that there are ways around the pocket veto, and so it doesn't necessarily get in the way of future attempts to pass legislation.

2. B

Opponents of the pocket veto argue that the practice gives the president too much power at the expense of Congress, as evidenced by the phrases *accumulation of power in the executive relative to the legislative branch* and *imbalance of federal powers through the subservience of the legislative bodies to the whims of the executive branch* at the end of paragraph 1 and beginning of paragraph 2, respectively. **(B)** most closely fits this prediction. **(A)** is Out of Scope: neither the author nor opponents and proponents of the veto suggest that the president exercises legislative authority. The question is whether the veto serves as a check on or threat to this authority within the legislative branch. **(C)** is a Distortion. While opponents of the veto think that it grants the president too much power, there's no evidence that their solution would be to allow an override of the pocket veto. Indeed, the opponents *may* have this opinion, but do not *have* to based on what we've read; on Test Day, that makes for an incorrect answer. Finally, **(D)** is an Opposite. If opponents of the pocket veto dislike that type of veto, they'd hate an absolute veto. The author says in paragraph 1 that opponents of the veto already criticize it as absolute.

3. D

The author considers *blatantly political use of the pocket veto* to be justified because *circumspection and deliberation are the more valued aspects of the lawmaking process*. The author is in favor of practices like the pocket veto that promote further discussion of proposed legislation. **(D)** summarizes the prediction and the reasoning in the passage. **(A)** and **(B)** are both Opposites. The author argues in the last paragraph that the pocket veto is justified even when blatantly political because it would allow more time for thought, which would contradict the assertion that it is unwise. Nowhere does the passage suggest that the president should be perceived as apolitical. Finally, **(C)** is Out of Scope. Although the author considers the pocket veto appropriate, presidential promises are not mentioned as a contributing factor.

4. D

First, decode the question stem. The strongest argument *against* the opponents of the pocket veto should be an argument in *favor* of the pocket veto (or at least a refutation of the opponents' conclusion). Paragraph 3 states that the pocket veto is a beneficial component of the system of checks and balances, and paragraph 4 says that the pocket veto is supported because of its historical use as such. As part of the system of checks and balances, the pocket veto allows the president to stall popular legislation in favor of additional deliberation. **(D)** matches this argument in favor of the pocket veto. **(A)** is an Opposite: the author says that a regular veto probably would have been overridden in the cases in question; however, this does not make **(B)** correct. This is a Faulty Use of Detail; the cases given are merely examples for what the author subsequently introduces as the *fundamental disagreement over the pocket veto*—fundamental being an Extreme Author keyword. This question asks for the strongest argument; this single piece of evidence is not nearly as strong as the idea presented in **(D)**. Finally, **(C)** is an Opposite. The author argues that the president sometimes does use the veto to override popular legislation; there's no reason to believe he wouldn't in this case also.

5. C

The *"Imperial Presidency"* is mentioned in paragraphs 1 and 3 as a derisive term used by opponents of the pocket veto. The author would not use this term, though. The author thinks that while an imperialistic presidency with absolute powers would be a bad thing, the current presidency isn't like that because it operates under Constitutional procedures—in fact, the author describes *shared responsibility between the president and Congress*. **(C)** reinforces this sentiment in the specific context of the pocket veto. **(A)** is Out of Scope; the author refers to the phenomenon of the *"Imperial Presidency"* only to show that the current system isn't as bad as opponents of the pocket veto suggest; the author does not think that the *"Imperial Presidency"*

currently exists in America, so it can't actively be a threat. **(B)** is a Distortion in that the author suggests that this type of presidency is similar to the *prerogative of the English kings*, but there's no suggestion that the *present* English system is like this. Finally, **(D)** is an Opposite; if the author doesn't think that the current presidency is imperial, then the *"Imperial Presidency"* can't be the cause of pocket vetoes.

Passage 2 (Questions 6–10)

Sample Passage Outline

P1. Expected population drop after a project may lead to poor infrastructure during construction

P2. Why money is scarce: taxing patterns, government jurisdictions

P3. Creation of boomtown: quest for energy, large population influx into small community

P4. Negative characteristics of boomtowns; most distinguishing = accelerated growth

P5. Negative social effects of development and effects on project

Goal: To describe the rapid growth and infrastructure problems of boomtowns

6. B

This is a Scattered Detail question and, as such, will require us to eliminate three answers that are supported by the passage to answer it. **(B)** is the only statement not suggested in the passage as a cause for lack of services. Although paragraph 5 mentions resentment between old-timers and persons brought to the community by the boom, there's no compelling reason why the lack of enthusiasm from long-time residents would lead to a shortage of schools, housing, or other services often missing from boomtowns. The other answer choices can all be found elsewhere in the passage. The loss of residents after construction finishes, **(A)**, is mentioned in paragraph 1. Taxation issues, such as those caused by an energy project's location outside the borders of an incorporated city, **(C)**, and delays in the onset of property taxes, **(D)**, are discussed in paragraph 2.

7. A
According to the author, public services are lacking in boomtowns primarily because of a lack of public infrastructure. Boomtown communities don't have the time or money to build the public support system needed to serve a large— but temporary—population. **(A)** correctly notes that such infrastructure, if in place, would improve public services in boomtowns. **(B)** is a Distortion. While taxation patterns and government jurisdictions are discussed in paragraph 2, a decrease in funding of any sort would almost certainly lead to a reduction (or have no effect on) public services, not improve them. **(C)** is also a Distortion. The author doesn't suggest that tax programs aren't well enforced—only that their structure prevents much tax revenue from being collected when it's needed. Finally, **(D)** is an Opposite. The author argues that problems arise when services don't rise to the short-term level needed by the town. When services stay at the level needed in the long term, the short-term residents are underserved.

8. B
The author discusses the relationship between boomtown woes and traditional taxation policies in paragraph 2. The author unhappily notes that *some entities may suffer the impact of development without being able to tax it*. **(B)** reflects the author's concern over the limitations of existing taxing programs. **(A)** is a Distortion. Although the author thinks that the inefficient taxation is a problem, there's no hint of outrage, which is Extreme. **(C)** is Out of Scope as there is nothing to suggest that the author is astonished by— or in disbelief of—the taxation programs. Finally, **(D)** is an Opposite. The author's criticism of the taxation issues faced by boomtowns implies a desire for change, so the tone is anything but complacent.

9. D
The author argues in paragraph 5 that boomtowns often suffer from divisions between the long-time and the short-term residents and that these divisions end up hurting everyone. **(D)** paraphrases this statement. The other choices are all Distortions. For **(A)**, although the long-time residents do suffer, there's no suggestion in the passage that they suffer *more* than new residents. For **(B)**, the author suggests in paragraph 5 that these problems are *result[s] of boomtown living*, but doesn't suggest that these problems specifically would be lessened with greater planning or consideration. Finally, for **(C)**, although the author mentions worker turnover in paragraph 5 and says that it's due to an undesirable community, it's not necessarily true that simply acknowledging long-time resident complaints would by itself improve the community enough to prevent turnover.

10. A
At the beginning of paragraph 5, the author cites a conclusion from studies. The conclusions are summarized (for example, *police and fire protection lag far behind population growth*), and then the author discusses the implications of those conclusions (*the fact remains that all parties suffer*). **(A)** is the only answer choice that parallels this structure: fact and follow-up discussion. All of the other answer choices are Out of Scope; because the opening line of the paragraph is retrospective, objective, empirical evidence, it cannot be classified as a prediction, point of view, or proposal.

4

Outlining the Passage

4: Outlining the Passage

In This Chapter

4.1 The Kaplan Method for CARS Passages   **60**
    Scan   61
    Read   62
    Label   63
    Reflect   64

4.2 Reverse-Engineering the Author's Outline   **64**
    Noteboard Strategy   65
    What to Label   66
    How to Highlight   66

4.3 Practicing the Strategy   **68**
    Sample Passage   68
    Analysis   70
    Example Outlines   73

Concept and Strategy Summary   **76**

Worked Example   **78**

Introduction

> **LEARNING GOALS**
>
> After Chapter 4, you will be able to:
>
> - Attack passages by applying the Kaplan Method for CARS Passages: Scan, Read, Label, and Reflect
> - Standardize your scratch work and highlighting
> - Summarize key points of an MCAT passage by constructing an outline

Great writers work with a plan. If you've ever had to construct an outline for homework as part of a larger essay assignment, your teacher was trying to impart to you this piece of wisdom. By forcing you to clarify your thesis and to articulate in just a few words what each paragraph is supposed to accomplish, outlining your paper typically makes writing itself much more streamlined, a simple matter of filling in the details to support your main points. On Test Day, your task will be to apply this insight by working backward, getting inside the mind of the author to reverse-engineer an outline that he could have used to compose the passage. As you'll soon discover, the process of constructing a Passage Outline sufficiently prepares you to answer the accompanying questions and serves as a valuable resource for efficiently navigating the text to score points quickly.

The plan for this chapter is as follows: first, we take a look at the four-step Kaplan Method for Passages called **Outlining**. The same method applies to science passages as well, but we'll examine how each of the steps is applied specifically in the *Critical Analysis and Reasoning Skills* (CARS) section. In the subsequent section, we discuss the hands-on approach to use on Test Day to construct your Outline with maximal efficiency, given the computer-based test format and limited space on your noteboard booklet. We conclude the chapter with an opportunity for you to practice the Kaplan Method for CARS Passages using a full-size passage and to evaluate your performance by comparing your process and results to our examples.

4.1 The Kaplan Method for CARS Passages

Great test takers also work with a plan! Figure 4.1 shows the Kaplan Method for Passages. We recommend following these four steps whenever approaching an MCAT passage.

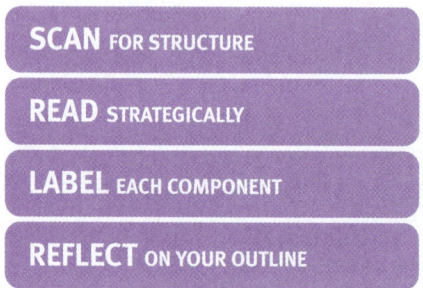

Figure 4.1. The Kaplan Method for Passages

While this general strategy is universal, it can be refined to fit the section that you're working on. As you've probably already discovered, there are several important differences between passages from the three science sections and those found in CARS. Most obvious is that the latter contain considerably more words, as well as having no images to break up the monotony of text. Under the surface, CARS passages are much more variable, both in their range of topics and their diversity of writing styles. Moreover, unlike the science passages, which are nearly always impartial, CARS passages are usually written by authors who take sides and express their opinions, although not always in a straightforward manner.

To account for these essential differences, the Kaplan Method for CARS Passages can be refined, as shown in Figure 4.2.

4: Outlining the Passage

SCAN FOR STRUCTURE
- Look for the big picture
- Assess the relative difficulty
- Decide to read *now* or *later*

READ STRATEGICALLY
- Read for organization—Relation keywords
- Read for perspective—Author keywords
- Read for reasoning—Logic keywords

LABEL EACH COMPONENT
- Briefly Outline the function of each paragraph
- Tag author opinions and alternative voices
- Note where to find evidence and refutations

REFLECT ON YOUR OUTLINE
- Ask: informative or persuasive?
- Choose a purpose verb
- Record the author's Goal

Figure 4.2. The Kaplan Method for Passages in CARS

Note: The Kaplan Method for CARS Passages, as well as the Kaplan Method for CARS Questions, CARS Question Types, and Wrong Answer Pathologies are included as tear-out sheets in the back of this book.

SCAN

The purpose of the **Scan** step is to give you an idea of what you're getting into, to allow you to decide whether this is the passage that you want to work on at this point in time. Scanning should only take a matter of seconds and is finished when you can answer the question: *Now or later?* If your answer is *now*, you'll continue with the other steps in the method. Should you instead decide *later*, you may wish to take a few seconds to jot down the topic of the passage and the number of its first question for reference on your noteboard, then move on to Scan the next passage.

Scanning passages is trickier in CARS than in the other sections, given the lack of figures with helpful captions—or anything besides paragraphs of text. Your best bet is to Scan for words that stand out due to capitalization, italics, quotation marks, parentheses, or any other distinctive textual features. When in doubt, just read the first (and possibly last) sentences of each paragraph until you start to get a sense of the big picture: the field of study, topic or theme, and rhetorical features like the author's purpose for writing. Remember that you don't need to read sentences word for word while Scanning: you're just trying to get enough information to assess the difficulty.

MCAT Critical Analysis and Reasoning Skills

The *now* or *later* decision ultimately comes down to your individual talents as a test taker. If you find that, say, technical philosophy passages take you a long time to complete with little payoff, then you'll want to save any such passages until the end. Why waste precious minutes on a difficult passage if filling in guesses would result in a similar score on that passage? Remember to pay attention not only to the subject matter but also to the difficulty of the language used. Puzzling vocabulary, convoluted sentences, and long-winded paragraphs can make even the most interesting topics frustrating to read. With enough practice, making determinations of difficulty will come quickly, and the small amount of time you invest in Scanning will pay off in points through better section management.

READ

Reading strategically means paying attention to keywords and the four modes of reading discussed in the preceding chapter: *content*, *organization*, *perspective*, and *reasoning*. Look beyond the content **buzzwords** to see how **Relation keywords** connect the different ideas in the text, how **Author keywords** offer glimpses of the writer's intentions, and how **Logic keywords** reveal the passage's arguments. If you don't yet feel comfortable identifying the different types of keywords, the analysis of the sample passage at the end of this chapter offers many examples that demonstrate how to recognize them.

As you read through each paragraph, you must resist the urge to reread the text excessively: you don't need to understand every single word and clause so long as you get a sense of the general direction of the author's discussion. A good rule of thumb is never to read a sentence more than twice during the Read step. If you don't understand a phrase or sentence the first time through, check whether a Relation keyword starts the next sentence—especially a **Similarity keyword**. If you see a Similarity keyword, it suggests that the author will continue with the same train of thought; often, the author will use simpler text to clarify the point made in preceding convoluted or complex sentences. Even if you feel lost after a second reading of the phrase or sentence, keep moving forward. You can always come back to that text for additional rereading if a question actually asks about it and points are on the line.

Above all, aim to figure out how a paragraph's sentences work together and what specific role that paragraph plays in the larger whole of the passage. The sooner you can figure out the author's overall purpose for writing, the easier it will be to see how each paragraph is supposed to function. That's why it's so helpful during the Scan step to look for those big-picture rhetorical elements (author, audience, message, goal, and context).

Bridge
Make sure to familiarize yourself with the different types of keywords and their functions. Keywords help you anticipate the path an author will take through the passage, and they allow you stay a step ahead of the author's thought process. Chapter 3 of *MCAT CARS Review* discusses the different classifications of keywords.

MCAT Expertise
Even if you don't understand a phrase or sentence after rereading it once, keep moving forward! Many MCAT test takers get caught up in complicated sentences and lose valuable time. If an author truly cares about a point she is making in a convoluted or complex sentence, the same point will likely return again in the passage—and usually in easier phrasing!

LABEL

The only components a CARS passage contains are paragraphs of text, so you'll be jotting down a brief description for each paragraph based on the results of your strategic reading during the **Label** step. Because you are the only person who needs to be able to decipher your Outline, you should strive to be efficient by using symbols, abbreviations, and other shorthand. Although Outlining will be a time-consuming process when you start practicing, you'll soon find a balance between being quick and being thorough. Like the Scan step, Labeling is a time investment. But because it will save you a considerable amount of effort when it comes to researching the questions, you'll find it becomes a net time *gain* once you've mastered Outlining.

Whenever possible, you should Label a paragraph in light of the four modes of reading discussed in Chapter 3 of *MCAT CARS Review*: content, organization, perspective, and reasoning. Not only should your Label reveal the content of that paragraph, but it should also give a sense of the paragraph's function within the larger structure of the passage, that is, the author's purpose for the particular paragraph. Make note in your Label of strong author opinions, as well as whenever an alternative perspective is being spoken for. Finally, pay special attention to the use of supporting or challenging evidence—many paragraphs function simply to strengthen or weaken a particular claim. Examples of each of these considerations will be offered later in this chapter.

It's essential to note that Reading and Labeling happen simultaneously. You should not start reading a new paragraph until you have some kind of Label for the one that you've just finished. Typically, it's worth reading the entire paragraph first before writing anything down because that tends to minimize the need for modifying your Label if the paragraph takes an unexpected turn. However, if the paragraph is especially long or dense, you might want to start jotting down a few words as soon as you get an idea of what's going on. Then, once you reach the end of the paragraph, just be sure your Label captures everything you feel is important before continuing.

It is worth saying that a Label in your Outline is not simply a set of notes. Writing out long lists of notes reduces the utility of your Outline because there is far too much detail for it to serve as a quick reference or guide to the passage as a whole. Further, taking notes usually accompanies a focus on content alone, rather than considering the larger function of the paragraph. Instead, a Label should be a brief summary of the paragraph as a whole, simultaneously encompassing the content, organization, perspective, and reasoning within that paragraph.

MCAT Expertise

Should I look at the questions before reading the passage? Because every passage is accompanied by five to seven challenging questions, we typically recommend that you don't read any of them before you've created your Passage Outline. That said, some students like to preview the questions at the very beginning to get a better sense of what to emphasize when reading. If you wish to try this alternative approach, take extra care in doing so. Above all, be sure you preview only the question stems and none of the answer choices because wrong options often promote misreadings of the text, which could end up costing you a lot of points. In addition, don't devote too much mental energy to remembering the stems or interrupt your reading by jumping to answer one of them; you are best served by concentrating on the passage to minimize rereading time.

Key Concept

Outlining, like any new strategy, will slow you down at first. But, with practice and internalization of the strategy, you will see that Outlining actually leads to an improvement in timing because you spend significantly less time researching answers to questions.

Real World

Creating Labels for a Passage Outline is similar to identifying the anatomy of the passage. We want to know not only what each part is but also how each part is connected and how each part relates to the greater whole.

Key Concept

The Goal of the passage is the author's purpose for writing it. In the Kaplan Method for CARS Passages, you write this Goal statement at the end of your Outline, always beginning with an infinitive verb such as *to explain*, *to argue*, or *to criticize*.

REFLECT

With each paragraph now Read and Labeled, all that remains is to pause briefly to **Reflect** upon the whole passage and articulate the author's overall **Goal** for writing, as described in Chapter 2 of *MCAT CARS Review*. Briefly look over the Labels in your Outline to get a sense of the overall arc of the passage. Did the author follow the same train of thought throughout the passage (perhaps a description of a social policy problem and a selection of potential solutions), or did he suddenly change the focus (perhaps describing the life of an individual composer and then criticizing the classification of musical genres throughout all of music history)? Because the authors of passages used in CARS are always trying to *do* something with their text, you should always begin a Goal statement with an action word, an infinitive verb (the form of a verb starting with *to*, such as *to describe*, *to analyze*, *to rebut*, and so on). The nouns and other parts of speech that follow will fill in the subject matter and the particular focus of the author's discussion. Occasionally, an author might have multiple major Goals, in which case it's okay to use more than one verb—just be sure to use *at least* one.

In the discussion of rhetoric in Chapter 2 of *MCAT CARS Review*, we noted that the purpose of most CARS passages will fall into one of two categories: **informative** or **persuasive**. When deciding on the Goal, consider where the author falls on the spectrum between taking a side and trying to remain impartial. This evaluation can help you to narrow down the appropriate purpose verb. Authors who are more informative can be said *to discuss* or *describe* a topic or, if going into greater depth, *to explain* or *analyze* it. If dealing with multiple ideas, the author might *compare* or *contrast* them. On the other hand, when authors are more persuasive, they may *support* or *advocate* a specific view, but many on the CARS section will simply *challenge*, *critique*, or *rebut* a position without offering any positive alternative. Using one or two of the example verbs considered here will work for most CARS passages, but feel free to choose others as appropriate.

As a final note, you don't actually have to wait until completing the passage to fill in the Goal. If you are able to discover the author's purpose during your Scan, or in one of the early paragraphs, feel free to jot it down. Just be sure that when you finish the passage, you take a moment to Reflect and make sure that the Goal was in fact what you originally suspected. All things considered, the final step should take no more than about 30 seconds, and potentially quite a bit less if you managed to identify the Goal early.

Bridge

In addition to clarifying the overall purpose of the author, the Goal statement is also useful for answering Main Idea questions, as described in Chapter 9 of *MCAT CARS Review*. The purpose verb is important too; many wrong answers in Main Idea questions will have a purpose verb with the wrong charge (*to advocate* vs. *to rebut*) or that is too strong or too weak (*to prove* vs. *to support* vs. *to describe*).

4.2 Reverse-Engineering the Author's Outline

Now that you're familiar with each of the steps of the Kaplan Method for CARS Passages, you're probably still wondering what precisely you should be doing on Test Day. In this section, we discuss how to optimize your noteboard usage, what to include in your Outline, and which portions of the passage are worth highlighting on the screen. After considering these practical concerns, you will be able to practice the method on your own until it becomes second nature for you to enter the minds of the authors of CARS passages and reconstruct their Outlines.

NOTEBOARD STRATEGY

On Test Day you will be provided with a noteboard booklet and wet-erase marker for scratch work. As you practice, keep scratch paper handy so that you can refine your technique. Because you won't be able to bring any personal items into the testing room, don't get attached to a "lucky" pen or pencil during your practice!

During the Scan step, you can begin to construct your Outline. Each paragraph should be numbered using brief notation like *P4*. Be sure to include a spot at the end for the author's Goal; because this will always start with an infinitive verb, you can write the word *to* as a reminder. For instance, you could set up your noteboard for a six-paragraph passage as follows:

P1.

P2.

P3.

P4.

P5.

P6.

Goal: To

This is the format that Kaplan employs in all of our example Outlines whenever we provide an explanation for a passage. Labeling becomes a matter of just filling in the blanks as you read—and the limited space will force you to be concise. Remember that abbreviation is your friend: just make sure that you're capable of understanding your own shorthand.

The challenge in constructing an effective Outline is to include the important information without writing too much. As a rule of thumb, a Label should be 5 to 7 words in a relatively simple passage and perhaps 10 to 12 words in a high-complexity passage. Labels should rarely, if ever, exceed 15 words or one line of text (keeping in mind that most students find it useful to fit two or even three columns onto their noteboard). You may find it helpful first to construct a complete sentence in your head that captures everything discussed below and afterward just jot down a few words and symbols that will jog your memory.

MCAT Expertise

Become a two-handed test taker. Most students use the same hand for both handwriting and using the computer mouse. Even though it may not seem like it takes long to switch between pencil and mouse, those precious seconds add up in a section with 53 questions. Thus, practice using the computer mouse with your non-writing hand. Not only will this allow you to highlight and Outline simultaneously, but you'll also find it saves you a surprising amount of time when dealing with calculations in the science sections.

WHAT TO LABEL

Once again, the four modes of reading from Chapter 3 of *MCAT CARS Review* are worth recalling. In this case, they serve as guidelines for the kind of material to include in each Label. In terms of **content**, the key ideas of each paragraph are worth jotting down, although you should stop short of trying to paraphrase every sentence. Instead, the important informational content is what is common to most if not all of the sentences in the paragraph, or the recurring theme that sets this paragraph apart from the others.

In addition to identifying the key ideas, your Outline is ultimately designed to reveal the **organization** of the passage, specifically how each paragraph functions within the larger whole. When Labeling a paragraph, think about what the paragraph is supposed to accomplish and the author's objective for placing it in the text. While many paragraphs function simply to convey information, more often than not, the content serves additional purposes, such as raising questions or problems that the author regards as significant, providing essential background for understanding an idea, supporting a claim made in the passage, or challenging a counterargument posed against the author's thesis.

The **perspectives** in the passage should be reflected in your Outline whenever the author has a strong opinion or in those cases when other voices speak. To Label a view as belonging to the writer of the passage, use *Author:* (or an abbreviation like *Auth:* or *Au:*) before the relevant text. Some expert test takers may use quotes (" and ") to indicate the author's opinion; if you choose to use this strategy, avoid using quotes elsewhere in your Outline. Similarly, use colons with other proper names to indicate alternative views. An easy way to denote positive and negative attitudes reflected in the passage is by using the shorthand of plus and minus signs. So, for instance, you could put "+ + +" next to an idea the author really likes. Or, if the author is ambivalent, having mixed positive and negative feelings about an issue addressed in one paragraph, you could include in your Label *(Auth ±)*. Some students prefer to use emoticons like smiley faces for the same effect.

Finally, because so many questions concern the **reasoning** of the passage, you want to pay special attention to supporting and challenging evidence as you read and be sure to identify its location in your Outline. Because many paragraphs exist simply to bolster an argument or object to one, you may already be taking into account the logic of the text when Labeling the paragraph's function. When support or challenge is not the focus of the paragraph, its presence may still be worth noting in your Outline within parentheses or after a dash.

HOW TO HIGHLIGHT

A final consideration is the use of the highlighting feature, which allows you to mark up the passage as it appears on the computer screen. Note that highlighting is never an adequate substitute for Outlining, but it becomes an extremely effective

MCAT Expertise

Do not rely on highlighting for all of the key information in a passage, but use it as a complement to your Passage Outline.

complement to the Kaplan Method for CARS Passages when used sparingly. Remember that highlighting works through contrast: if you highlight most of the text, the unmarked portions would actually stand out more. Consequently, strive to adhere to the following guidelines when using this feature.

Click, Drag, Release, and Select

On Test Day, you'll be able to highlight by holding down the left mouse button and dragging the cursor. When the left button is released, the text will have a gray background. At this point, you can highlight the selected text by either clicking on the word "Highlight" in the top-left corner of the screen (assuming "highlight," and not "remove highlight" is selected in the nearby drop-down menu) or by using the keyboard shortcut (alt + H). If you highlight a phrase, don't obsess about capturing it exactly: if you omit a few letters or grab an extra word or two, you'll still accomplish your intended purpose of calling attention to that specific part of the text. This sounds silly, but many test takers will lose valuable seconds making sure to highlight precisely. Never highlight anything longer than one line of text at once.

Find the First Occurrence

It's generally useful to identify when terms are introduced by an author, especially when the author provides an overt definition or offers essential background information. However, with recurring themes, it would take far too much time to highlight every instance the term appears. Scanning will often give you an idea of what these major ideas will be, but if you find a term that you didn't initially realize was important being repeated, just jump back to highlight where the term was defined before continuing with your reading.

Proper Nouns and Numbers

Names and other proper nouns, as well as dates and other numbers, have a tendency to show up in question stems as clues. Consequently, it's typically worth your time to highlight them as they appear, or at least, in accordance with the preceding guideline, at their first mention. You'll want to be especially careful with numbers, making sure that any answer choice that includes a time or other numerical value is consistent with the information you highlighted in the passage. The only possible "math error" you could make on the CARS section would be mixing up the numbers provided by the passage; highlighting numerical values can prevent you from committing such an easy-to-avoid mistake.

Keywords

Our final piece of advice may seem somewhat counterintuitive given the importance of keywords for reading, but as a general rule, you should not highlight most of the keywords that you encounter in a passage. While it's essential to recognize them as you read in order to appreciate the deeper levels of meaning in the text and devise a better Outline, there is little value in taking the time to highlight

> **MCAT Expertise**
> On Test Day, highlighting will be retained until you remove it (even if you navigate away from a passage). You will be able to highlight the passage, question, and answer choices.

them. Remember that highlighting is valuable only insofar as it makes it easier to research the passage and find the information you need to answer questions correctly. When researching a question, you typically won't pay much attention to the keywords until *after* you've isolated the relevant text.

The notable exceptions to this rule are those cases in which the keywords do reveal important aspects of the passage's structure that would facilitate answering the questions. Logic keywords are the most likely to fill this role. As an alternative to marking down the location of supporting or challenging claims in your Outline, you can instead opt to Label only the conclusion on your noteboard and just highlight the important Evidence or Refutation keywords. The other major case in which you should highlight keywords occurs with relations in longer paragraphs. If a lengthy paragraph is split by a major internal contrast or a series of steps, highlighting the Difference or Sequence keywords can illuminate the organization of the text on screen.

4.3 Practicing the Strategy

Treat the following passage as an opportunity to practice the Kaplan Method for CARS Passages. While your decision will ultimately be to do the passage *now*, you can still Scan to determine the topic and relative difficulty. Grab a sheet of scratch paper so that as you read strategically by watching out for keywords, you can Label each paragraph to create your Outline. Finally, don't forget to take a moment to Reflect and identify the author's Goal.

After attempting the method on your own with the sample passage, read the subsequent analysis section to see how closely you followed our recommended approach; then compare your Outline with the following example Outlines. While there is not one single correct way to Outline a passage, not all Outlines are created equal: the true test of an Outline will be whether it helps you to answer the questions correctly and efficiently. However, because we won't presently be looking at any questions with this passage, you can at least use our examples as guidelines.

SAMPLE PASSAGE

Can we truly know anything with certainty? Since the dawn of the so-called "Early Modern" era in Western philosophy, this question has preoccupied both skeptics and their critics. Perhaps the most noteworthy challenger of the certainty-rejecting skeptics is René Descartes, who constructs in his seminal *Meditations* an elaborate argument that purports to ground all human knowledge on the indubitability of one's own existence. However, in his attempts, Descartes actually bolsters the case for skepticism. Indeed, later thinkers even cast doubt on the supposed surety of self-existence.

4: Outlining the Passage

Dissatisfied with the dogmatism of his scholastic forebears, Descartes sought to clear away all the questionable but typically unquestioned "truths" handed down to him and his contemporaries, expecting that anything that remained after an onslaught of radical doubt would have to be known with certainty. Demolishing accepted opinions one by one would require volumes, so instead Descartes examines the basic categories of belief, rejecting any kind for which he can find plausible reasons for doubt.

The first *Meditation* begins the process by considering empirical knowledge, what is learned from experience by means of our senses. Descartes's most powerful argument relies upon the impossibility of discerning waking consciousness from sleeping. Who has not dreamt of "waking up" while still asleep? When I awaken, how do I know the "reality" around me is not just another layer of illusion, a dream-within-a-dream-within-a-dream? Thus, Descartes concludes that all knowledge that derives from sensation cannot be certain.

With the certainty of the *a posteriori* now eradicated, the *Meditations* turn to the *a priori*, knowledge that is independent of experience, such as mathematics and logic. This proves a more difficult task, so Descartes must introduce the possibility of a Great Deceiver, a malevolent being with godlike powers who deludes us at every turn. If I cannot prove that such an entity does not exist, then "how do I know that I am not deceived every time that I add two and three, or count the sides of a square, or judge of things yet simpler, if anything simpler can be imagined?" Readers of Orwell's *Nineteen Eighty-Four* might have an easier time of imagining this, recalling that Winston Smith under torture genuinely comes to believe that $2 + 2 = 5$.

If even arithmetic can be cast into doubt, then how could anything be known for sure? Descartes provides an answer in the second *Meditation*. Even if the Deceiver tricks me about everything else, he cannot delude me about my own existence: "Let him deceive me as much as he will, he can never cause me to be nothing so long as I think that I am something." Of course, this "I" that exists for certain does not include the physical body, which may just be an illusion, but is simply the thinking self or mind. Even so, Descartes builds on this proposition in the remainder of the *Meditations*, arguing first for the existence of a benevolent God who would not deceive us about anything perceived "clearly and distinctly," subsequently enabling him to announce an enormous ontological undertaking.

But is Descartes's foundation really so certain? Is it not possible to doubt the existence of one's own mind? In the *Genealogy of Morals*, Friedrich Nietzsche raises the possibility of an even more radical skepticism. Though

Nietzsche uses the example of lightning, the same point might be clearer to English speakers with the statement "it is raining." Although "it" seems to suggest some agent independent of the action, the phrase simply means that raining is happening. Rather than saying with certainty that "I think," perhaps Descartes should have merely concluded that "thinking happens."

ANALYSIS

In conducting a Scan of the passage, a lot jumps out even in the first paragraph: the question mark in the first sentence, *Can we truly know anything with certainty?*; the quotes in *so-called "Early Modern" era in Western philosophy*; and the capitalization and italicization in *René Descartes, who constructs in his seminal* Meditations. Skimming over the first lines of the remaining paragraphs yields more references to Descartes and the *Meditations*, with a little bit of Nietzsche thrown in at the very end. This is clearly a philosophy passage, predominantly focusing on something that the thinker Descartes wrote in his book called the *Meditations*. Many students find philosophy to be the single most challenging field that appears in the CARS section, but *now* is the perfect time to see that philosophy passages are not nearly so daunting as their reputation would suggest.

Paragraph 1

When a passage opens with a question, chances are that the answer to this question will be the focus of the text. An opening question like this is not a standard rhetorical question because it does not have one uncontroversial answer that the audience could be safely assumed to adopt. It will require further work to figure out how the author would answer, and finding this answer should be our primary goal at this point. Continuing, the next sentences offer some important historical context: there are two schools of thought on the question, the *certainty-rejecting skeptics* on the one hand and their *critics* such as Descartes, who is described as *[p]erhaps the most noteworthy challenger*. This last phrase is rich in Author keywords: *noteworthy* is Positive, but the Extreme keyword *most* is canceled out by the Moderating keyword *perhaps*, so it's clear that the author's view of Descartes is not just flagrantly positive.

A critical stance toward Descartes is confirmed by the concluding sentences of the paragraph. *However* is a Difference keyword that is meant to contrast two things: Descartes's *attempts* (a Negative keyword because the suggestion is that something that is attempted could always fail) in the *Meditations* and what is *actually* (an Extreme keyword, indicating a strong author opinion) the effect, namely to *bolster* or support the opposite position—that of the skeptics. The final sentence with the mention of *supposed surety* further confirms that the author questions the soundness of Descartes's pro-certainty conclusion.

4: Outlining the Passage

Paragraph 2

The next paragraph may seem unexpected because it lacks any kind of transitioning Relation keyword. The two sentences that constitute this paragraph both have Descartes as the subject, and the use of the past tense suggests that the author is providing historical context. The upshot is that Descartes began his quest for certainty somewhat counterintuitively: by rejecting all the suspect beliefs of his education. The key idea that untangles this paradox, however, is that *anything that remained after an onslaught of radical doubt would have to be known with certainty*. Given that *certainty* is the subject of the opening question, you'll want to note additional instances of it, such as this one.

Paragraph 3

You find a more palpable transition here, with the Sequence keywords *begins the process* signaling that more steps are to be expected. For the first step, though, you'll want to be clear on what is distinctive, and this opening sentence is useful by providing a technical term and its definition: *empirical knowledge, what is learned from experience by means of our senses*. The words that follow, *Descartes's most powerful argument relies upon,* should strike you like a bolt of a lightning, containing Logic keywords as well as strongly Positive Author keywords. What follows *relies upon* must be evidence. The two questions that follow function differently than does the opening question: they are in fact rhetorical and are supposed to support this *powerful argument*. The final sentence's *Thus, Descartes concludes* reveals a conclusion that the author seems sympathetic to but is ultimately attributed to Descartes.

Paragraph 4

Remembering the Sequence keyword from the third paragraph, we now see the next one: *turn to*. The author follows this with another technical definition: *the* a priori, *knowledge that is independent of experience, such as mathematics and logic*. Now, the sequence begins to emerge: first doubt experiential beliefs, then doubt mathematical ones. While the author was fond of Descartes's previous *powerful argument*, here we find *This proves a more difficult task*, a comparison that reveals the author likely regards the second argument less highly. This paragraph otherwise seems to follow the model of the preceding paragraph, with a rhetorical question, this time in quotation marks (suggesting Descartes's voice), being used for support. The reference to Orwell at the end might be baffling to you, but keep in mind that the testmakers do not expect you to know anything about a work of fiction unless it's explicitly stated in the passage, so all you need to understand is that it's being used in a supporting role.

Paragraph 5

Does the question at the start of this paragraph look familiar? It's almost like the one at the beginning, and you're told immediately about where *Descartes provides an answer*. Before reading another word, you should already recognize that this is going to be important because after the sequence of the prior two paragraphs, the author is finally returning to the point. The next sentence contains a colon followed by a quote; the colon is functioning as an equals sign, suggesting that the first statement made in the author's voice is the same as the second statement made in Descartes's. That means if you don't understand one, you can rely upon the other. The basic idea is that you can't be fooled into believing you exist when you actually don't because if you didn't exist in the first place, there would be no "you" to fool.

So how does the author feel about this? Before, we noted sympathy toward the skeptical position and even some slight hostility toward Descartes. The author's attitude is affirmed when the limits of Descartes's claims are discussed in the following sentence: that the self whose existence was supposedly proven is *simply*, or merely, the mind. The following *Even so* establishes a difference between what the author thinks and the text that follows: the author is suggesting that, despite this narrow limitation in the meaning of the self, Descartes still claims to ground all of knowledge on the existence of his own mind.

The end of paragraph 5 presents a rare but delightful poetic device called **alliteration** in which the author repeats the same phoneme over and over again in stressed syllables: e**n**abling him to a**nn**ounce a**n** e**n**ormous o**n**tological u**n**dertaking. While this is an infrequently used device, it serves an important purpose. Much as your eye would be drawn to a bright or shiny object, your ear (or internal representation of your voice, at least) should perk up when hearing this phrase. While you should certainly not *expect* to see alliteration on Test Day, authors may use other rhetoric, poetic, or literary devices to similarly "decorate" their writing. This hints that the author particularly cares about the phrase, and it's likely to be important to the overall Goal of the passage. Note that the poetic definition of alliteration may differ from ones you've heard before—the repeated phoneme need not be located at the beginning of words but rather can occur at stressed syllables in the phrase.

Paragraph 6

Now we reach rhetorical questions once again—*really* and *not* should be dead giveaways that the author wants you to answer by saying *no, Descartes's foundation is not really so certain* and *yes, it is possible to doubt the existence of your mind*. How is this possible? Well, the author assures you that Nietzsche does it with *an even more radical skepticism* than the one that Descartes used to clear away uncertain beliefs. To be more radical, it must call into question what Descartes does not. While not revealing precisely what Nietzsche says, the author

> **Bridge**
> A phoneme is the smallest unit of speech sounds, like "b," "t," or "sh." Note that every major word in this alliterative phrase contains a stressed "n" sound. Phonology describes the phonemes in a given language, and it is one of the five basic components of language, as discussed in Chapter 4 of *MCAT Behavioral Sciences Review*.

offers an analogy with familiar language used to describe the weather. If "it" can rain without *it* referring to anything that's doing the raining, then why can't the same apply to thinking? The author ends with a bit of an understatement: *perhaps Descartes should have merely concluded that "thinking happens."* The Moderating keywords suggest once again that an author of a CARS passage is hedging and making a weaker claim than is probably necessary.

Goal

Having examined each paragraph, all that remains now is to return to the big picture and Reflect on the Goal of the passage. The author is not at all neutral, attempting to persuade the audience to adopt beliefs about the validity of Descartes's arguments. The tone of the passage, moreover, is overwhelmingly critical, as reflected by all the Negative Author keywords noted above. This suggests that a purpose verb like *challenge* or *critique* would be most apt in formulating the Goal.

To be more specific, the author is challenging Descartes's answer to the question posed at the very beginning. Descartes thinks that knowledge can be known for certain and offers his reasoning in the *Meditations*, which the author recounts in several paragraphs but all for the sake of undermining it. Descartes's primary reason for believing that knowledge can be sure, according to the author, is that the existence of the thinking self cannot be doubted. But in the final paragraph, the author takes reasons from Nietzsche to suggest that it *can* in fact be doubted. You may not have realized its importance at first, but this is precisely what the author said would happen in the last sentence of the opening paragraph—*Indeed, later thinkers even cast doubt on the supposed surety of self-existence.* Authors like this one have a tendency, despite the difficulty often presented by the abstract concepts they discuss, to be straightforward about exactly what they'll do in a passage.

EXAMPLE OUTLINES

Each of the following examples is a perfectly legitimate way of Outlining the sample passage; collectively, these examples serve as varying reflections of the analysis above. If your Outline looks like any of them, or something of a hybrid, then you're off to a great start! As noted by the names provided, the major difference is in the extent of material offered. While most students will want to strive for a balance between being thorough and being concise, a bare-bones Outline has the advantage of taking the least time to construct.

The Fully Fleshed-Out Model

P1. Is certain knowledge possible? Skeptics say no; critics like Descartes say yes (Author likes skeptics)

P2. Looking for certainty, Descartes rejects any established category of belief that can be doubted

P3. D's 1st *Meditation*: beliefs from experience doubtable—supported by waking/dreaming argument

P4. More *Meditations*: beliefs independent of experience doubtable—supported by Great Deceiver idea

P5. D's 2nd *Meditation*: the thinking self is undoubtable—allows for certainty (assuming benevolent God)

P6. Author: Nietzsche undermines Descartes with reasons to doubt certainty of thinking self's existence

Goal: To explain Descartes's arguments for certain knowledge and criticize him for insufficient skepticism

The Bare-Bones Version

P1. Auth: skeptics (nothing = certain) > Descartes (self = certain)

P2. D rejects belief categories if doubtable

P3. D's argument against experience

P4. D's arg. against math/logic

P5. D's arg. for certainty of thinking self

P6. Nietzsche's criticism of ↑ (Auth +)

Goal: To challenge D's argument for certainty

A Balance of Breadth and Brevity

P1. Descartes argues for certain knowledge, grounded in self-existence (but Author prefers skepticism)

P2. D isolates certain truths by rejecting any category of doubtable belief

P3. D: can't distinguish waking/dreaming → experiential beliefs = doubtable

P4. D: Great Deceiver might fool us → math/logical beliefs = doubtable

P5. D: thinking self can't be doubted + nondeceptive God → certain knowledge

P6. Author/Nietzsche: D is wrong, thinking could happen without a self

Goal: To present and critique Descartes's argument for grounding knowledge on certain self-existence

Conclusion

With the right strategic approach, CARS passages become much more manageable. Scanning the passage to discern its difficulty can allow you to manage your time more effectively, reordering the section to work best for you. Strategic Reading means paying attention to keywords and the additional levels of meaning that they unveil. Labeling each paragraph lets you create a brief Outline, a guide for efficiently navigating the text as the questions require. Finally, Reflecting on the entire text to discern the author's Goal enhances your overall understanding as you discover how the parts function within the larger whole of the passage. The following two chapters should aid you in further developing your understanding of perhaps the most crucial aspect of CARS passages: the logic of their arguments.

MCAT Critical Analysis and Reasoning Skills

CONCEPT AND STRATEGY SUMMARY

The Kaplan Method for CARS Passages
- **Scan** for structure
  - Look for the big picture
  - Assess the relative difficulty
  - Decide to read *now or later*
- **Read** strategically
  - Read for organization—Relation keywords
  - Read for perspective—Author keywords
  - Read for reasoning—Logic keywords
- **Label** each component
  - Briefly Outline the function of each paragraph
  - Tag author opinions and alternative voices
  - Note where to find evidence and refutations
- **Reflect** on your Outline
  - Ask: informative or persuasive?
  - Choose a purpose verb
  - Record the author's **Goal**

Reverse-Engineering the Author's Outline
- Begin to construct your Outline before even reading the passage, numbering each paragraph and providing a space for the author's Goal.
- A good Label is brief and covers all four modes of reading (as appropriate).
  - While you should find the average amount of depth that works best for you, most Labels can be written with 5 to 7 words in simple passages and 10 to 12 words in high-complexity passages.
  - The key content from a paragraph should be noted in the Label.
  - The Labels together form a **Passage Outline**, which shows the organization of the passage and how each paragraph functions within the larger whole.
  - Include the author's (and others') opinions in your Outline using abbreviations or quotation marks. Positive and negative attitudes (and ambivalence) can be illustrated with plus and minus signs.
  - Major aspects of logic (conclusion, evidence, and refutation) can be noted in the Label.

4: Outlining the Passage

- Use on-screen highlighting sparingly.
  - Highlight the first occurrence of major terms used in the passage or where the term is defined.
  - Highlight names and proper nouns, dates, and other numbers.
  - Only highlight keywords that reveal important aspects of a passage's structure, like Logic keywords and Sequence keywords.

Practicing the Strategy

- There is no single correct way to Outline a passage, but not all Outlines are created equal.
- Your best Outline is one that will help you answer the questions correctly and efficiently.

MCAT Critical Analysis and Reasoning Skills

WORKED EXAMPLE

Use the Worked Example below, in tandem with the subsequent practice passages, to internalize and apply the strategies described in this chapter. The Worked Example matches the specifications and style of a typical MCAT *Critical Analysis and Reasoning Skills* (CARS) passage.

Passage	Analysis
One of the more well-known female writers to adopt a pen name was George Sand, born Aurore Dupin, who became one of the most prolific and admired French authors during the nineteenth century. The true identity of George Sand did not remain a secret for long, for the author used this name in her everyday life, and close friends commonly referred to her as "George."	We are introduced to Aurore Dupin, who wrote under the pen name George Sand. It wasn't much of a secret, considering her friends knew her as George and she used the name in her everyday life. We don't know exactly where the author is going with this—is the focus on the author or the pen name? For now, keep the Label short and sweet: **P1.** Dupin = Sand, pen name famous
Most portraits of the author as an adult are simply *George Sand* and make no reference to her given name. Her son, too, adopted this new last name, even though association with his famous author–mother did not bring him any obvious benefits. Given the name "George Sand" is radically different from Aurore Dupin's birth name, many readers have wondered how the author formulated her masculine pen name.	We have two additional examples of the use of the pen name, but finally a question is posed: why and how did Dupin choose such a masculine pen name? Expect the author to speculate on some possible reasons. We Label this paragraph as: **P2.** More examples, question—why Dupin chose masculine pen name?
At least two possible answers spring to mind. The first, as indicated in Curtis Cate's biography *George Sand*, is that the pseudonym arose from a collaboration with her first lover, Jules Sandeau, with whom she co-authored several articles as well as a full-length novel entitled *Rose et Blanche*. On the advice of their publisher, the authors signed this latter work under the name "J. Sand." Once Aurore's writing began to overshadow that of Jules, she decided to sign her solo works as "Georges Sand," which eventually became simply "George Sand." Because her own literary output was a great success, she quickly became known by this name and began to use her pen name on a daily basis.	Identify the *two possible answers* that the author proposes. The first theory is given by Curtis Cate: Dupin chose the name to separate her solo works from the collaborative works with her former lover written under "J. Sand." The second theory has not yet been mentioned, so we need to keep an eye out for it in the remainder of the passage. A Label for this paragraph might be: **P3.** Theory 1: Cate—new name separates solo work from collaborative works with former lover

78

4: Outlining the Passage

Passage	Analysis
By continuing to use the name initially assigned to collaborative writings with her lover, perhaps Aurore hoped to maintain their connection. Perhaps she fondly remembered their time together and wished to have a permanent reminder. Or perhaps she simply realized that it would be much more expedient to continue to write under a name which was already familiar to her audience thanks to their joint works.	The word *perhaps* is used a lot here regarding Dupin's feeling for Jules and the decision to retain the name "Sand," so we have more speculation. And yet, still no mention of the second theory—keep your eyes open for it. We might Label this as: **P4.** Possible reasons for keeping "Sand" in Theory 1
Given that George Sand began writing under this masculine name around the same time as she began to roam around Paris in pants and a jacket—typically male clothing—it is not hard to understand why she chose a masculine pseudonym because, like her clothes, this male identity gave her more freedom of expression, both literally and figuratively. And once she became known as a successful author under this name, there was no reason to change. Writing under a false name allowed her to distance parts of her character—her roles as wife, mother, and lover—from the creative and literary parts that formed her role as an author. Using a male name set her apart and added to her persona as an unusual and fascinating woman. In the end, the reason why she chose this particular pen name is not nearly as important as the vast quantity of writing—articles, letters, novels, plays—that forms her legacy to the field of French literature.	A male identity in clothing and pen name gave Sand more freedom of expression, allowed her to distance herself from other roles, and added interest to her persona. But in the end, the author tells us that the reason for the name is less important than her overall high-quality writing. There's still no mention of that second theory, but there's also only one paragraph left—it must be there. We might Label this paragraph as: **P5.** Reasons for choosing a male name; Auth: Dupin is great, regardless

MCAT Critical Analysis and Reasoning Skills

Passage	Analysis
The name could have a more symbolic meaning as well which would give more deserved credit to the author herself. Taking each letter of "SAND" as an allusion to names, places, or people from Aurore's life, this name can be seen as a representation of Aurore's childhood and early married life. Even if George created the name, however, she was well aware of the similarity to her lover's name, and was equally aware that many readers would make this connection. As an intelligent and perceptive woman, she recognized that such an association with a male author would help to validate her early writing career before she had succeeded in establishing her own reputation as a talented and publishable author.	Finally, the second theory! "SAND" may be an acronym constructed from references to different aspects of Dupin's life. However, *even if* this theory were true, choosing a name similar to a famous lover would give her some credibility as an author early in her career. Tying all this new information into a Label, we have: **P6.** Theory #2: acronym; more reasons to choose name (validation)

Here's a sample Outline and Goal for this passage:

P1. Dupin = Sand, pen name famous

P2. More examples, question—why Dupin chose masculine pen name?

P3. Theory 1: Cate—new name separates solo work from collaborative works with former lover

P4. Possible reasons for keeping "Sand" in Theory 1

P5. Reasons for choosing a male name; Auth: Dupin is great, regardless

P6. Theory #2: acronym; more reasons to choose name (validation)

Goal: To discuss two theories for why Aurore Dupin chose the pen name George Sand

4: Outlining the Passage

Question	Analysis
1. According to the passage, the following were all possible reasons for George Sand to create a pseudonym EXCEPT:	The phrase *according to the passage* indicates that this is a Detail question; the word *EXCEPT* makes it a Scattered Detail question, and we are looking for what is not mentioned. We will need to reference the passage and use process of elimination to eliminate all of the answer choices that were mentioned in the text.
A. she began publishing collaborative works with Jules Sandeau.	We can eliminate **(A)** because it was mentioned in paragraph 3 where the author discussed Cate's theory.
B. her new name reflected important parts of her life.	**(B)** was described in the final paragraph where the author mentioned that each of the letters in SAND could represent allusions to parts of Dupin's life.
C. she was not able to publish any works under her own given name.	**(C)** is the correct answer because the author never mentioned that Dupin was unable to publish under her given name. While the author states that her pseudonym may have helped validate her writing and provided greater freedom of expression, no direct evidence is given that Dupin could not publish any works under her name.
D. the works published under her pen name sold well.	**(D)** was mentioned in paragraph 3 where Dupin had published previously with her lover under the name of J. Sand, adopted the name George Sand, and published successfully.

MCAT Critical Analysis and Reasoning Skills

Question	Analysis
2. With which of the following statements would the author most likely agree?	Because this question is asking us what the author would most likely agree with, we are looking at an Inference question. We will want to make a general framework for what we expect the correct answer to look like based on what the author has discussed. The pen name George Sand had multiple possible origins—a connection to an old lover or a more symbolic meaning—but regardless of the origin, the pen name gave Dupin more freedom of expression as discussed in paragraph 5.
A. Aurore Dupin should have written works under her own name once the secret of her pseudonym was revealed.	**(A)** is an Opposite. The author stated that *once she became known as a successful author under this name, there was no reason to change*, so the author would not think she should have changed her name.
B. By writing under a pseudonym, George Sand created for herself a new identity that allowed her to transcend the limitations of society.	**(B)** closely matches the prediction, highlighting the freedom Dupin's pseudonym gave her. This is the correct answer.
C. George Sand owed her early success to her partner, Jules Sandeau.	**(C)** is a Distortion that misconstrues Dupin and Sandeau's literary output. While Dupin began her writing by collaborating with Sandeau, the author does not imply that her early success was due to him.
D. The choice of a masculine pseudonym was restrictive for George Sand and forced her to live as a man throughout her life.	**(D)** is both a Distortion and an Opposite. Although the pseudonym George Sand certainly was masculine, it did not force Dupin to live as a man; she wore pants and a jacket of her own free will. Furthermore, the masculine pseudonym was freeing, not restrictive.

4: Outlining the Passage

	Question	Analysis
3.	The author mentions Curtis Cate in order to:	This is asking us why the author mentioned a particular example, so it is a Function question. We will need to refer back to the passage where Curtis Cate is discussed. According to the Passage Outline, the author refers to Cate in paragraph 3. In this paragraph, the author mentions that there are at least two possible reasons why Dupin chose the name George Sand and introduces Cate's theory as one possible reason. The author does not say anything negative about Cate's theory, so we expect the tone of the correct answer to be neutral.
A.	refute his claims about the reason for Aurore Dupin's choice of a male pseudonym.	**(A)** is a Distortion because the author never refutes Cate's claims. While the author does present another possible reason, the author does not imply that Cate's theory is invalid.
B.	provide support for a plausible explanation of the creation of Aurore Dupin's pseudonym.	**(B)** matches the prediction, both in terms of content (*creation of Aurore Dupin's pseudonym*) and tone (*provide support*). This is the correct answer.
C.	advocate the reason for Aurore Dupin's pseudonym as presented in this particular biography.	**(C)** is also a Distortion because the author does not suggest that Cate's theory is the best; the author just mentions it as one of many possible reasons. *Advocate* is too strong a word to describe the author's writing.
D.	show that biographers do not always write accurately about their subjects.	**(D)** is Out of Scope because the author never makes any claims about the accuracy of biographers on their subjects.

MCAT Critical Analysis and Reasoning Skills

Question	Analysis
4. According to the passage, there was widespread use of the pseudonym George Sand. Based on the points the author brings up, which of the following is NOT proof of this widespread use?	The question stem contains the phrase *according to the passage*, which would usually indicate a Detail question. However, the question is asking us to identify evidence (related to logic), so this is actually a Function question. The word NOT means this is a Scattered Function question. We will need to use the passage to find the pieces of evidence and eliminate them. It may seem like a bit of digging, but we can use our Passage Outline to narrow down the search area. The author is discussing the widespread use of the pseudonym George Sand mainly in paragraphs 1 and 2.
A. Members of her family used part of her pseudonym for themselves.	**(A)** can be eliminated because the author stated that Dupin's son adopted the last name Sand in paragraph 2.
B. Aurore Dupin's lovers and close friends called her "George."	**(B)** can also be eliminated because the author mentioned that many close friends referred to her as "George" in the last sentence of paragraph 1.
C. Portraitists and the general public knew her predominantly by her pen name.	**(C)** was also mentioned in paragraph 2 where the author stated that portraits of the author went by the name George Sand and didn't mention the author's given name.
D. Early book reviews of her works never referred to her given name.	By process of elimination, the answer must be **(D)**. The author did not mention early reviews of her books, so **(D)** was not used as evidence.

4: Outlining the Passage

Question	Analysis
5. Regardless of the facts set forth in the passage, the author implies that the second possible reason for George Sand's pen name is:	The word *implies* in the question stem indicates that this is another Inference question—specifically, an Implication question. The first theory, Curtis Cate's, predominates in paragraphs 3 and 4 whereas the second theory, SAND as an acronym, is presented in paragraph 6. The author introduced this second possible theory as one that *g[a]ve more deserved credit* to Dupin, suggesting that the author somewhat agrees with this theory.
A. appealing because it demonstrates Dupin's creativity and independence.	The word *appealing* in (A) matches the author's positive opinion of this theory. The use of an acronym (and a pseudonym at all) would be good evidence of *Dupin's creativity and independence*, which the author already praised in paragraph 5. This is the correct answer.
B. equally plausible as the first reason even though it has no relevance to the Dupin's family.	(B) is an Opposite wrong answer choice as the second reason states that the name originated from *names, places, or people* in Dupin's life.
C. too sentimental for such a rational and innovative writer.	(C) is a Distortion because the author believes that Dupin, as an *intelligent and perceptive woman, would* recognize the benefit of using the pseudonym—rather than shying away because it seemed *too sentimental*.
D. based on reading she did during her childhood and early married life.	(D) is Out of Scope. The author does not mention any references to books that Dupin read.

MCAT Critical Analysis and Reasoning Skills

Question	Analysis
6. According to information put forth by the author within the confines of the passage, George Sand's male pen name and her choice of clothing are related because:	This phrase *according to* hints that this is a Detail question, so our first step is to locate the relevant information. The author mentioned the male clothing and masculine pen name in paragraph 3, saying that *it is not hard to understand why she chose a masculine pseudonym because, like her clothes, this male identity gave her more freedom of expression*. In other words, she was able to overcome limitations by taking on a male identity.
A. both acknowledge her strong masculine side.	(A) is a Distortion because although Dupin wore more masculine clothing, the author never suggested Dupin herself had a *strong masculine side*.
B. both provide evidence of her androgyny.	(B) is also a Distortion as there is nothing to suggest that she was androgynous (having equal characteristics of both masculinity and femininity).
C. both freed her from stereotypical female constraints.	(C) matches closely with the prediction and is the correct answer; Dupin's adoption of a male identity (both pseudonym and clothing) allowed her to overcome limitations based on her sex.
D. both permitted her to succeed in a patriarchal society.	(D) is a nuanced Distortion of the author's points. The adoption of a male identity helped Dupin overcome obstacles in her way, but did not itself lead to her success—her success is attributable to *establishing...her reputation as a talented and publishable author*, as stated at the end of paragraph 5.

Practice Questions

Passage 1 (Questions 1–5)

Because it impinges upon so much—from bilingual education, political correctness, and Afro-centered curricula to affirmative action and feminism—the current discussion on multiculturalism is essential to understanding Western academic culture today. Charles Taylor's account of the development of multiculturalism out of classical liberalism traces it through changing conceptions of what he terms "the politics of recognition."

Deft as his historical account may be, any analysis of the motivations for multiculturalism solely in terms of "recognition" must remain fundamentally incomplete. In his analysis are two central demands for recognition underlying classical liberal thought: the demand for the equal recognition of human dignity, and the demand for the recognition of and the respect for all human beings as independent, self-defining individuals. Multiculturalism, according to Taylor, rejects both of these ideals and their political application in an official "difference-blind" law (which focuses on what is the same in us all). Instead, it embraces laws and public institutions that recognize and even foster particularity—that cater to the well-being of specific groups. These two modes of politics, then, both having come to be based on the notion of equal respect, come into conflict.

Taylor acknowledges that it can be viewed as a betrayal of the liberal ideal of equality when the multiculturalist calls for a recognition of difference rather than similarity, and seeks special treatment for certain groups—such as aboriginal hunting privileges or the "distinct society" of Québec. However, he plausibly argues that to recognize only sameness is to fail to recognize much that is necessary for real "recognition" because we are all cultured individuals with personal histories and community ties. Still, Taylor does not stray far from classical liberalism, insisting that multiculturalism be able to "offer adequate safeguards for fundamental rights."

The more extreme forms of multiculturalism, which Taylor disavows, commit the crucial error of reducing all ethical and normative standards to mere instruments of power because in doing so any distinctly moral arguments for these positions become absurd. Though Taylor seems correct to reject this diminution, he's wrong to think that the "recognition" model alone can sufficiently account for the demands made by various minority groups for both the promotion of discrete cultural identities and the transformation of the dominant culture. What many in these groups desire is much more than mere recognition or approval: it is the power to more effectively and independently control their own destinies.

It's even become common to disdain the respect or solidarity professed by those in the dominant group in an attempt to consolidate separate cultural identities. How Taylor misses this fact is not clear because even his favorite example of Québec's distinct society presents a case in which the primary function of the demand for recognition is to acquire the power necessary for those within to maintain, promote, and even enforce their way of life. Taylor understands that the Québécois want more than to merely preserve their culture, or to have others appreciate it. They also want to create a dynamic, autonomous society in which future generations will participate as part of a common project. Unfortunately, he does not consider how this fact undercuts the notion of "recognition" as an adequate lens through which to view their project.

1. According to the passage, extreme multiculturalists make which of the following mistakes?

   A. They wrongly disdain the solidarity professed by those in the dominant group.
   B. They undercut their position by eliminating any moral arguments that they could make for their view.
   C. They overestimate the distinctness of their position from classical liberalism.
   D. They argue for their position on moral grounds, rather than pointing to the practical benefits of their views.

2. According to the argument posed by the author in the passage, multiculturalism may be seen as a betrayal of liberal ideals because:

   A. classical liberalism is not concerned with the well-being of minority groups.
   B. it abandons the demands for equality that characterize classical liberalism.
   C. a failure to recognize what is different about individuals can be a failure to fully recognize individuals.
   D. it is not capable of respecting diversity or offering adequate safeguards for basic rights.

3. The author's two references to the "distinct society" of Québec are primarily intended to:

   I. give an example of a multiculturalist demand.
   II. give an example for which Taylor's analysis is inadequate.
   III. give an example of a group for which special treatment is sought.

   A. I only
   B. III only
   C. I and II only
   D. I, II, and III

4. Which of the following can most reasonably be inferred from the passage about the author's attitude toward the two classical liberal ideals of equality?

   A. They are adequate for most contexts in which recognition is demanded.
   B. They do not safeguard fundamental rights for individuals in aboriginal groups.
   C. They reflect a disguised attempt by a privileged group to maintain its power over other groups.
   D. They reflect an impoverished conception of the individual person.

5. Based on the information provided in the passage, it would be most reasonable to expect the author to agree that Charles Taylor's "politics of recognition" model is:

   A. admirable, but flawed.
   B. of limited use.
   C. historically deft.
   D. an unmitigated failure.

Passage 2 (Questions 6–11)

Peter Gay's book, *The Education of the Senses*, re-examines Victorian bourgeois attitudes about sensuality and sexuality in an attempt to discredit the pervasive and negative view of the Victorian bourgeois as repressed and repressive people whose outward prim public appearances often hypocritically masked inner lascivious thoughts and private behaviors. One of the most interesting facets of Gay's study is his discussion of the necessary, yet taboo, issue of birth control during the latter part of the nineteenth century.

Gay points out that the very process of giving birth was dangerous to both the newborn and the mother—that most women suffered greatly during the birth, that many children and new mothers died within five years of a birth, and that many women approached the child-bearing process with trepidation although they believed that producing offspring was a woman's ultimate fulfillment. Advice or assistance from the medical profession—whether licensed doctors or self-trained midwives—was sorely lacking and inconsistent, hardly capable of reassuring the expectant mother and father as to the woman's and the baby's safety.

In fact, the medical profession itself was largely responsible for promulgating myths and rumors about the dangers of attempting to limit family size through use of some forms of birth control, regardless of the fact that so many women and children died each year due to complications of pregnancy or birth. Some medical, religious, and social experts did acknowledge that the continual cycle of birth was not only detrimental to the health of the mother but also could take a toll on the quality of life of the family because multiple children increased the financial burden and responsibility of the father. Furthermore, women were often caught in this cycle of pregnancy and childbirth well into their late 40s, greatly increasing the health risks and mortality rate of mother and child alike. It seemed appropriate and even necessary, then, to make efforts to limit the number of offspring in order to benefit the family unit and thus the greater good of the larger society itself.

Nonetheless, open discussion of birth control methods, both natural and device-assisted, was rare, even between a doctor and his patient. Most information was passed along by word-of-mouth, which inevitably led to a great deal of unchecked misinformation that was, at times, deadly. Gay maintains that a primary motivation for this reticence was deeply ingrained in the Victorian bourgeois mindset that emphasized the value of family and traditional roles and thus encouraged women to be productive—in the very literal, procreative sense. Though concerned husbands certainly did take steps to assure that their wives and families were not jeopardized by an overabundance of offspring, a widespread effort to limit family size was not firmly rooted in society until the advent of a strong women's movement, which did not make many real and meaningful strides in changing public attitudes and behaviors until the early twentieth century. Thus, Victorian bourgeois women were obliged to fulfill their societal role as child bearers despite very real fears about the toll this could take on their health and on the well-being of their family.

MCAT Critical Analysis and Reasoning Skills

6. Which of the following is NOT, according to the author, a reason that most Victorian women did not use any form of birth control?

   A. The medical profession did not offer any methods of birth control, but only vague suggestions for limiting the number of pregnancies a woman had during her lifetime.
   B. Although the risks to women were widely recognized, women were still held responsible for continuing the family line.
   C. Accurate information about safe forms of birth control was difficult to ascertain, even from doctors or midwives.
   D. Many medical experts were reluctant to discuss birth control with their patients because they themselves were not well informed about the options and consequences.

7. Which of the following, if true, would most seriously WEAKEN Gay's argument that deeply ingrained social attitudes were responsible for the lack of open discussion about birth control?

   A. The majority of Victorian women were satisfied with their role as wives and mothers.
   B. Limiting family size was inconsistent with bourgeois perceptions about social status.
   C. Bourgeois Victorians believed the ideal family consisted of fewer than five children.
   D. Many Victorian physicians actually knew that family planning would improve women's overall health.

8. The passage suggests that which of the following was/were commonly associated with childbirth during the Victorian era?

   I. Death of the child
   II. The need for additional family income
   III. Appreciation of the mother's suffering

   A. I only
   B. I and II only
   C. I and III only
   D. II and III only

9. Which of the following general theories would be most consistent with Gay's arguments as presented in the passage?

   A. Examination of sociohistorical context offers the modern historian little useful information about prevailing attitudes and behaviors of a certain time period.
   B. Understanding the psychological basis for social actions can help to explain apparently contradictory behavioral patterns.
   C. Social historians can best analyze past cultures by applying modern theories to the work of earlier critics.
   D. Critics should not attempt to rework prior studies of social classes because they cannot properly understand the historical context of their predecessors.

10. The passage suggests that Peter Gay is LEAST likely to agree with which of the following statements?

    A. Social critics should occasionally re-examine existing beliefs to make sure that they are appropriate.
    B. The prevailing views of Victorian bourgeois society were based on accurate sociocultural perceptions.
    C. The women's movement of the early twentieth century helped to bring the issue of birth control into the public consciousness.
    D. Sometimes, common mindsets must be altered before real social change can be accomplished.

11. Based on the passage, which of the following does the author consider was the most important factor contributing to the high danger associated with Victorian childbirth?

    A. False and misleading information about the risks of childbirth
    B. Unsanitary conditions in the birthing rooms
    C. Successive cycles of pregnancy and childbirth
    D. Oppressive social attitudes that forced women to procreate

Explanations to Practice Questions

Passage 1 (Questions 1–5)

Sample Passage Outline

P1. Introduction to Taylor's analysis of multiculturalism; "politics of recognition"

P2. Taylor: classical liberalism's two demands for equal recognition; multiculturalism rejects these (recognizes differences instead)

P3. Multiculturalism seems anti-liberal by recognizing difference; Taylor: "recognition" also depends on differences, but safeguard rights

P4. Extreme multiculturalism too narrow; Auth: Taylor's "recognition" model insufficient to explain minority groups' goals

P5. Author's criticism of Taylor: groups' desires are more than just recognition, gives example (Québécois)

Goal: To critique Charles Taylor's analysis of multiculturalism and its focus on recognizing diversity

1. B

This is a Detail question looking for the mistakes that extreme multiculturalism makes. According to our Passage Outline, extreme multiculturalism is discussed in paragraph 4. The author states that extreme multiculturalists *commit the crucial error of reducing all…standards to mere instruments of power…and in doing so make moral justifications absurd*. We can make a prediction that the extreme form eliminates moral arguments. **(B)** best matches this prediction. **(A)** describes a mistake some minority groups make, as described in the beginning of the fifth paragraph, but doesn't refer to extreme multiculturalism. **(C)** could potentially describe Taylor, given the author's claim that *Taylor does not stray far from classical liberalism*. Still, this choice does not describe extreme multiculturalists. Finally, **(D)** is an Opposite—indeed, the author argues that moral arguments are lost in extreme multiculturalism.

2. B

This is another Detail question, so we will need to use our Passage Outline to guide us where to go. The Label for paragraph 3 indicates that multiculturalism differs from liberalism by recognizing differences—so what does liberalism recognize? As noted in the first sentence of this paragraph, the liberal ideal that is betrayed is equality. **(B)** matches cleanly with this prediction. **(A)** Distorts the author's argument and makes it Extreme. Although liberalism does not concentrate on minority groups to the same extent as multiculturalism, there is no evidence that suggests that is has no concern at all for minority groups. **(C)** is the justification Taylor gives for multiculturalism, not the reason why multiculturalism betrays liberal ideals. Finally, **(D)** is an Opposite as Taylor insists that multiculturalism *offer[s] adequate safeguards for fundamental rights*.

3. D

Question 3 is a Scattered Function question because we are asked to interpret the author's motives for including a particular example. The example of Québec is mentioned in paragraphs 3 and 5, so we will need to analyze the content in which it is mentioned. In paragraph 3, the author notes that the *"distinct society"* of Québec sought special treatment, so Statement III is true. We can eliminate **(A)** and **(C)** because they don't have III. The Québécois are mentioned again in paragraph 5 as an example of where Taylor's analysis is inadequate, matching what is said in Statement II. **(D)** is the only answer choice that has both II and III, and thus must be the correct answer. Statement I is also true because the Québécois are mentioned as an example of one society's *multiculturalist calls for recognition*.

4. D

This is an Inference question as it directly asks what can be inferred from the passage. According to our Passage Outline, liberal ideals were discussed in paragraphs 2 and 3. The author states that Taylor acknowledged the deviation from liberal ideals and that Taylor's justification is *plausibl[e]*. According to Taylor, concentrating on sameness—the focus of liberalism—*fail[s] to recognize much that is necessary for real "recognition."* The liberal ideals fail to fully recognize the individual, which makes **(A)** incorrect and finds a match in **(D)**. **(B)** is a Distorted version of the concern some may have about multiculturalism's rejection of equality and does not describe the author's attitude toward the liberal ideals. Finally, **(C)** is Out of Scope because the topic of privileged groups maintaining power over others is never discussed. The dominant group was briefly mentioned in paragraph 5, but not in the context of maintaining power.

5. B

This is another Inference question in which we need to identify the author's feelings toward Taylor's model. We can use the Goal in our Passage Outline to make a preliminary prediction: because the author is critiquing Taylor's ideas, the author believes there are some flaws. More specifically, the author introduces Taylor's historical account as *deft*, but claims that his *analysis ... [is] fundamentally incomplete*. **(B)** is the best match for this prediction. **(A)** and **(C)** are subtle wrong answers. For the former, the author certainly does believe that Taylor's ideas are flawed, and the language does not suggest admiration. For the latter, while the author describes Taylor's historical account as *deft*, he or she has a different opinion about the model itself—it is incomplete. Finally, **(D)** is a Distortion as it is Extreme: while the author views Taylor's model as incomplete, he or she does not believe that it is an absolute failure.

Passage 2 (Questions 6–11)

Sample Passage Outline

P1. Introduces Gay's re-examination of Victorian bourgeois attitudes; will focus on birth control

P2. Describes dangers to women and their babies; Victorian women: "childbirth = ultimate fulfillment"

P3. Doctors exacerbated the misinformation that women received, but recognized some issues with constant cycle of birth

P4. Discussion of birth control remained hidden in Victorian society—women had to fulfill their roles (until women's movement)

Goal: To examine Peter Gay's study on Victorian bourgeois views on birth control

6. A

This is a Scattered Detail question asking for what is *not* stated as a reason women did not use birth control. We will need to use the process of elimination along with our Passage Outline to eliminate the answers discussed in the passage. Starting with **(A)**, the medical profession's attitude about birth control is mentioned in paragraph 3. Doctors did not provide even vague advice for limiting family size, but rather *promulgat[ed] myths and rumors about the dangers of attempting to limit family size*. Thus, **(A)** is the correct answer. **(B)** is mentioned at the end of paragraph 4. **(C)** is mentioned in detail in paragraph 2 and briefly mentioned again in paragraph 4. **(D)** is discussed in paragraph 3 and again at the beginning of paragraph 4.

7. C

This is a Strengthen–Weaken (Beyond the Passage) question focusing on what would *weaken* the author's argument. The author argues that deeply engrained social attitudes were responsible for the lack of open discussion of birth control. We will need to consider what the social attitudes were at the time, as the right answer will contradict them. The author stated that women *believed that producing offspring was a woman's ultimate fulfillment* and that *a widespread effort to limit family size was not firmly rooted in society.* A prediction for the answer would likely be something that implies that women felt that they had a different role or wanted to limit the size of the family. **(C)** best matches this prediction. **(A)** is consistent with Gay, reflecting on the idea that Victorian women *believed that producing offspring was a woman's ultimate fulfillment*, and does not significantly weaken Gay's argument. **(B)** would strengthen the author's argument, as it gives an additional reason why women would not aim to limit family size—and would likely not discuss birth control openly, by extension. Finally, **(D)** would also strengthen the author's argument because it removes one of the alternative explanations why birth control was not discussed—if doctors actually knew that limiting family size was beneficial, and yet still did not discuss it openly, then it increases the likelihood that prevailing social attitudes limited discussions of birth control.

8. B

Because the question asks what the passage *suggests*, this is an Inference question; the Roman numeral answer choices make it a Scattered Inference question. We will need to evaluate the Roman numerals one by one to arrive at the answer. Starting with Statement I, the process of childbirth and the consequences of it were mentioned in paragraph 2. The author stated that many *children...died within five years*, so Statement I is true, and we can eliminate **(D)**. For Statement II, the author mentioned in paragraph 3 that additional children *increased the financial burden and responsibility of the father*. This suggests that the family would need a larger income to support a larger family, so Statement II is also true. When we look at the answer choices, there is only one answer that has both Statements I and II, so **(B)** is the correct answer. For reference, Statement III is false because while some husbands were concerned about their wives during childbirth, it was not widespread.

9. B

This is an Apply question in which we need to use our understanding of the author's argument to determine what statement would be most consistent with it. The author mentioned the problems associated with childbirth in the second paragraph and the lack of action by medical professionals in paragraph 3. The author then explained that the reason birth control was not discussed was due to the current social mindset in which women were expected to be productive. This argument structure is most consistent with **(B)**. **(A)** and **(D)** are Opposites; Gay uses sociohistorical context and reworks prior studies about Victorian bourgeois attitudes—so these statements would be inconsistent with Gay's argument in the passage. **(C)** is Out of Scope as Gay does not apply modern theories to earlier critics, but rather re-evaluates information and research on the Victorian bourgeoisie themselves.

10. B

This is an Apply question (specifically of the Response subtype) asking us what Gay would be least likely to agree with. Gay's study was a re-examination of the stereotypical beliefs about Victorian bourgeois attitudes and behaviors with regard to sexuality. We can predict that if Gay found it necessary to re-evaluate the information of the Victorian bourgeois, he must have found the existing studies and opinions to be inaccurate. **(B)** best matches this prediction. **(A)** is an Opposite because this is exactly what Gay did—so he would definitely agree with re-examining existing beliefs. **(C)** and **(D)** both reflect on the end of the passage, where the author states that widespread attempts to limit family size did not occur until a strong women's movement made meaningful strides in changing public attitudes and behaviors.

11. A

This is another Inference question asking what the author thought was the most important factor contributing to the danger of childbirth in the Victorian era. While the author did not explicitly state which factor he or she thought was the most important, we can look at how the argument was structured for insight. The predominant focus of the passage, which plays a role in paragraphs 2, 3, and 4, is the misinformation from the medical field. This matches closely with **(A)**. **(B)** is Out of Scope as sanitation was never discussed. **(C)** is indeed mentioned in the passage but merely as an example of one of the few things doctors recognized was problematic—it is too narrow and restricted of an answer when compared to the author's overall argument in the passage. Finally, **(D)** is Extreme; social attitudes certainly encouraged women to procreate, but the word *forced* is simply too strong.

5
Dissecting Arguments

5: Dissecting Arguments

In This Chapter

5.1 Domains of Discourse	**98**	**5.4 Arguments: Conclusions and Evidence**	**103**
Potential Confusions	99	Counterarguments	105
Where Does Logic Fit?	100	Inferences	106
5.2 Concepts: The Basic Elements of Logic	**100**	Strengthening and Weakening	107
Relations of Ideas	101	**Concept and Strategy Summary**	**109**
5.3 Claims: The Bearers of Truth Value	**102**		
Consistency and Conflict	102		
Support	103		

Introduction

> **LEARNING GOALS**
>
> After Chapter 5, you will be able to:
>
> - Differentiate between the natural, textual, and conceptual domains of discourse
> - Explain the importance of concepts in logical arguments
> - Separate claims from concepts using their relative truth value
> - Distinguish evidence from conclusion within a logical argument
> - Recognize counterarguments, inferences, and assumptions within an argument

Imagine for a moment you wrongly believed that mitochondria were vital organs of the body, or that ligaments and tendons were what held molecules together. Even though the MCAT is ultimately a test of critical thinking, such fundamental anatomical inaccuracies would inevitably lead to misunderstood passages and missed points. The *Critical Analysis and Reasoning Skills* (CARS) section is not so different: it would be just as flawed to talk about the *truth of a concept* or the *falsity of an argument* as to describe the *kidney of a cell* or the *ribosome of an atom*. Fortunately for you, the structures employed in reasoning are *far less* complex than the intricate inner workings of human physiology, with its manifold tiers of organization and seemingly boundless cavalcade of molecular participants operating at every level. Consider this chapter your guide to the comparatively simple anatomy of arguments.

MCAT Critical Analysis and Reasoning Skills

We begin the chapter with a context-setting discussion, describing three domains of discourse that are essential to distinguish when working with CARS passages. The remainder of this chapter is devoted to defining a three-tiered logical hierarchy—a simplified model of reasoning designed to give you just enough logical acumen to think clearly about CARS passages. This explanation begins with an introduction to the lowest level, containing the elementary units of reasoning—concepts. Subsequently, we'll investigate claims, the only items that can properly be said to be true or false. At the highest tier we find arguments, which constitute the subject of our final section. Throughout the chapter, we will entertain a number of examples, but our prime specimen will be the philosophy passage from the previous chapter, ripe for dissection with its rich argumentative structure.

5.1 Domains of Discourse

When working with any academic text, as with all of the passages found in the CARS section, you will find it useful to keep distinct in your mind three levels that we call the **domains of discourse**: the natural, the textual, and the conceptual. While the language here may sound technical, the basic difference we're getting at is one that you likely already understand: things *vs.* words *vs.* ideas.

The domain of the **natural** refers to everything that can be found in the world around us. This encompasses both the objects that make up the physical world, from the simplest subatomic particles to the local galactic supercluster, and the events that occur within it, from the briefest motion of a photon to the entire lifespan of the universe. The word *natural* should be understood broadly here: it includes the distinctively "human" affairs studied by both the humanities and social sciences, like individual experiences, shared cultures, socioeconomic institutions, and historical civilizations; the "artificial" aspects of the world, such as technology and the arts; and even what might be called the "supernatural" or "metaphysical," discussed in a passage on religion, for instance.

The **textual** denotes the realm of words, sentences, and paragraphs—everything that directly faces you in an MCAT passage. It can be difficult to isolate the textual domain because language immediately points beyond itself—to objects, concepts, or perhaps both—on account of the associations that we are trained to make when learning to speak and to read. Nevertheless, it is essential to recognize that words have their own unique characteristics and structures independent of what they represent, including the rhetorical aspects of language, analyzed in Chapter 2 of *MCAT CARS Review*. Passages in literature and related disciplines tend to focus heavily on the textual domain.

> **Key Concept**
>
> The term *domains of discourse* is used to delimit three distinct types of discussion—the natural, the textual, and the conceptual—each of which has distinctive parts and relationships that must not be confused.

> **Bridge**
>
> In the study of language, the term *semantics* is used to describe the connection between a specific set of phonemes (language sounds) and a given meaning for a word. This is one of the five basic components of language, as discussed in Chapter 4 of *MCAT Behavioral Sciences Review*.

Lastly, **conceptual** discourse concerns the sphere of concepts, claims, and arguments, the primary focus of this chapter and the next. Much more will be said about it below, but for now, it might be helpful to think of the conceptual domain the same way you do about some scientific phenomena. In the natural world, you will find no truly ideal gases or thermodynamic systems with 100 percent efficiency, and yet we can still talk about their distinctive properties as abstract entities. To clarify: we're not trying to suggest that there's some mystical separate world of ideas existing independently of the natural world. Rather, we're saying that there are special considerations that apply uniquely to concepts, allowing us to treat them *as though* they were their own separate sphere of existence.

POTENTIAL CONFUSIONS

It's important to stress that these domains of discourse can never be entirely separated from one another and even share a number of analogous structures, as described in Chapter 6 of *MCAT CARS Review*. Nevertheless, being able to isolate the concerns of each in your mind will help to clarify your thinking—and not only when it comes to working on the CARS section. Some of the most common errors in judgment involve the merging of one domain of discourse with another. Let's consider examples of each possibility.

Confusing the natural and textual worlds occurs when one assumes that an author's description of the world automatically corresponds to the world as it exists. It's always important to remember that every author brings her own perspective to the text, with opinions and biases that tilt the presentation of information. This is why it's generally a good idea to question what you read and think about the author's motivations behind the text. Authors are human beings who, like you, have had a limited subset of experiences but who use that finite experience to make conclusions about reality. If you can get inside an author's head and imagine where she is coming from, you'll more easily distinguish objective facts from the author's subjective impressions.

The textual and conceptual worlds are perhaps the two most often interchanged for one another, as is reflected in the common occurrence of heated disputes over the meanings of words. Have you ever had a debate with another person about an abstract issue, only to discover after several minutes of tense back-and-forth that you were actually agreeing with one another all along, just using different words to express the same ideas? (If you haven't, it may just be the case that this occurred in the past without either of you realizing it!) Conversely, just because two people use the same word does not mean that they refer to the same idea, as in the use of politically charged words such as *freedom* and *democracy*—virtually everyone today seems to agree that these are worthy ideals, yet individual conceptions of them vary dramatically.

MCAT Critical Analysis and Reasoning Skills

Finally, misunderstandings between the natural and conceptual realms can be some of the most difficult to untangle. One example includes the confusion between justification and causation, which will be explained in Chapter 6 of *MCAT CARS Review*. An everyday example is the phenomenon of stereotyping, in which a person is pigeonholed into a category, then prejudicially treated as a member of a group, rather than as an individual with unique characteristics and values.

WHERE DOES LOGIC FIT?

Now that the importance of distinguishing these domains is becoming clear, you may be wondering what it has to do with the subject of these chapters: logic. Though we sometimes use the word more broadly to mean something like *structure*, in its more technical sense, **logic** is the formal study of argumentation or reasoning. This places it squarely within what we've identified as the conceptual domain.

> **Key Concept**
>
> Logic is the formal study of arguments. It falls into the conceptual domain of discourse.

As a branch of both mathematics and philosophy, logic is approached in a wide variety of ways, some of which are extremely advanced and well beyond the purview of the MCAT. While a complete course in formal logic might be useful for Test Day, it is certainly not essential. We deliberately employ a more simplified approach to logic that will not require an entire semester to master. The more technical aspects of our treatment will be handled in the following chapter, but the remainder of this chapter will focus on defining the key concepts in logic and how they interconnect. We stick to using terms that actually appear in CARS question stems, but note that the MCAT will not always use language quite as precisely as we do.

> **Bridge**
>
> As introduced in Chapter 2 of *MCAT CARS Review*, appeals to logic are called *logos*. These can be differentiated from appeals based on credibility, *ethos*, and appeals to one's emotions, *pathos*.

Within the study of logic, there are three nested levels. The most basic of these, which gives the conceptual domain its name, is the concept. Concepts relate together to make up claims, analogous to atoms making up molecules in the natural world or words making up sentences in texts. These claims, in turn, combine in arguments, like molecules forming cells or sentences composing paragraphs. At each level—concepts, claims, and arguments—there are distinct characteristics. We begin with the most basic of the elements of reasoning: concepts.

> **Key Concept**
>
> Concepts or ideas are the logical analogs of atoms, the basic units of the conceptual domain of discourse, which takes its name from them. A concept has a meaning or definition, but it is not by itself true or false.

5.2 Concepts: The Basic Elements of Logic

Also known by such words as *idea* or *notion*, a **concept** is the fundamental building block of logic. Concepts have **meanings**, but they are not true or false, properly speaking. Note that the word *meaning* can be ambiguous, but here we use it specifically to point out the **connotation** or **definition** of an idea; that is, the set of distinctive properties that make the idea what it is. Concepts can still have definitions, even if they don't refer specifically to anything in the natural world: while there is no such thing as a *unicorn*, the idea still carries the sense of a magical, white stallion crowned with a single, golden horn.

As suggested above in the discussion of the three domains, concepts should be distinguished both from the **terms** used to represent them, on the one hand, and from the natural objects or events that might be given as **examples** of them. For instance, the concept of a *table*, as in the piece of furniture, is separate from any actual table that you may presently be sitting at and even from all the tables in the world taken together. The concept is also distinct from the combination of letters *t–a–b–l–e*, which can represent any number of other meanings, such as a noun meaning *a graphical representation of data* or a verb meaning *to delay discussion of an issue*.

The authors of passages used in CARS will often use words in ways to which you're unaccustomed, taking a term that may be familiar but using it in a narrow, specialized sense. Watch out for explicit definitions, examples, and contextual clues to help you get a sense of an idea's meaning, particularly when specific words and phrases are frequently repeated. Be warned, though, that many authors will use multiple terms interchangeably to refer to the same underlying concept and that, even further, questions in CARS can employ synonyms and paraphrases of ideas in different wording than is used in the passage. In short, it will be essential on Test Day to look for concept-for-concept correspondences, not exact word-for-word matches.

> **MCAT Expertise**
>
> MCAT students often lose points because they search for answer choices that use exactly the same wording as the passage. It is not uncommon for the testmakers to write answer choices that reflect the same concept using different words (a different textual representation).

RELATIONS OF IDEAS

In Chapter 3 of *MCAT CARS Review*, Relation keywords were introduced as a tool for recognizing the organization within texts. Many Relation keywords are useful on a textual level, revealing how the author intends diverse sentences to be connected. However, in addition to revealing points of continuity or contrast within a text, Relation keywords will also tie ideas together in distinctive ways. In our description of Relation keywords, we pointed out that the most basic relationships are Similarity and Difference, but much greater specificity is possible, such as Opposition, Sequence, and Comparison. One other important relation to note is that between a whole and its parts—for instance, the *golden horn* is a concept that is also a part of the concept *unicorn*, which constitutes a larger whole.

In addition to relating to each other, concepts can also relate to parts of the other domains of discourse. On the one hand, a natural object or event might *exemplify* or *embody* a concept, meaning that the entity serves as an example of it. On the other hand, a term or word might *represent*, *designate*, or *refer to* a concept, meaning that the word or phrase is being used to stand in place for the concept. While these are not technically relations between ideas, you should still be aware of these interactions between the different realms of discourse because these words are used this way in CARS questions.

MCAT Critical Analysis and Reasoning Skills

5.3 Claims: The Bearers of Truth Value

What distinguishes a **claim** (also called an *assertion*, *statement*, *proposition*, *belief*, or *contention* on the MCAT) from a mere concept is **truth value**, the capacity to be either true or false. While claims may be quite complex (potentially consisting of numerous concepts related together in diverse ways), having a truth value requires only two parts at a minimum: a **subject** and a **predicate**, like the most important nouns and verbs in a sentence. Thus, *the Abominable Snowman* is not a claim but rather a subject without a predicate, making it a concept with meaning but no truth value. However, *the Abominable Snowman is eight feet tall*, *everybody loves the Abominable Snowman*, and even *the Abominable Snowman doesn't exist* are all claims because all are either true or false, in addition to being meaningful.

Just as it's essential to know the difference between a term and the concept that it denotes, so too should you be able to distinguish between the sentences of a text and the *claims* that underlie them. Again, authors often find many ways of making the same point, echoing a claim while varying the language that they use to sound less repetitive. Furthermore, CARS question stems and answer choices will contain their own clauses, which could depart dramatically from the words the author uses while still retaining the same sense. This is all the more reason for distinguishing between the textual (the words the author uses) and the conceptual (the ideas those words represent).

CONSISTENCY AND CONFLICT

Claims are important in CARS on account of the relationships that they share with one another, particularly those that concern truth value. These must be distinguished from the relationships mentioned above that concepts share, which are based on meaning, not truth. **Consistency** is the simplest relation between claims: if two or more statements can be true simultaneously, they are described with language like *consistent*, *compatible*, or *in agreement* on the exam. Bear in mind that this is a pretty minimal relationship. Most logically consistent claims have absolutely nothing to do with one another: *Though commonly perceived as overwhelmingly motivated by the struggle to end slavery, the American Civil War was in actuality the product of a hodgepodge of disparate political, economic, and cultural forces, among which abolitionism was but one prominent factor* is completely consistent with *I like cookies*.

Because being in agreement is "easy," the opposite relation is fairly extreme: when CARS passages and questions use terms like *inconsistent*, *contradictory*, or *conflicting*, they refer to a set of claims (typically just two) for which it is *impossible* that all be simultaneously true. Contradictory claims could be outright negations of one another, differing only in that one includes a word like *not* or a prefix like *un–*, or they could be incompatible in a more subtle way like *it is 80°F outside* and *it is snowing*.

Key Concept

Claims are the middlemen in the logical hierarchy, composed themselves of concepts and their relationships and, in turn, composing arguments. Claims consist of at least a subject and a predicate, and they have both meaning and truth value (the capacity to be true or false).

Bridge

Note that sometimes the MCAT will use consistency terms in conjunction with *most* or *least*, asking for the *claim that is most conflicting* or one that is *least consistent with the author's ideas*. In those cases when agreement is treated as a matter of degree, the terms are being used with meanings closer to *similar* and *dissimilar*. See the discussion of analogical reasoning in Chapter 6 of *MCAT CARS Review* for more on this topic.

SUPPORT

Because it is a factor in about half of all questions you'll encounter in CARS, there is no relationship more important to understand than **support**. This is a special type of connection that is unidirectional in nature, unlike the "two-way street" of consistency that treats claims as equal in status. A statement is *supporting* when its truth would increase the likelihood that the claim *supported* by it is also true. Thus, for one claim to support another, the two must also be consistent. At the same time, unless the argument is circular, support will not flow the other way.

An illustration of this latter point can be seen when considering the philosophy passage from Chapter 4 of *MCAT CARS Review*, when the author notes in the third paragraph that *Descartes concludes that all knowledge that derives from sensation cannot be certain* because of the *impossibility of discerning waking consciousness from sleeping*. In effect, the claim *dreaming cannot be distinguished from waking reality* supports the statement *empirical knowledge is uncertain*. If we were to ask why we should believe that waking can't be distinguished from sleeping, Descartes could not reasonably say *because empirical knowledge is uncertain* because that's what he was originally trying to convince us of in the first place!

> **Key Concept**
> Support is a unidirectional relationship between claims. If one claim supports another, the truth of the first would make the truth of the second more likely, but *not* vice-versa.

The opposite of support is the relationship sometimes known as a *refutation*, *challenge*, or *objection*. This will be addressed later in the discussion of counterarguments.

5.4 Arguments: Conclusions and Evidence

In support relationships, claims get special names. A mere contention becomes a supported **conclusion** as long as it's accompanied by supporting **evidence**. Evidence is like the friend who vouches for you to get into the VIP area at a nightclub (*Conclusions only!*): it offers reason to "trust" that the conclusion is what it claims to be. In other words, evidence is an answer to the question *Why should I believe that?* Together, the evidence, the conclusion, and the one-way connection between them (support) constitute the simplest form of an **argument**. This is what the example of support from the Descartes passage looks like in argumentative form:

- **Evidence**: It is impossible to tell the difference between dreaming and being awake.
- **Conclusion**: Knowledge gained from sense experience is inherently uncertain.

> **Key Concept**
> An argument is the combination of one claim, known as the conclusion, and one or more other assertions, known as the evidence, explicitly used to support it.

While one piece of evidence is sufficient to make a claim count as a conclusion, authors typically rely upon multiple sources of support, sometimes even providing additional reasons for believing the evidence provided. So, continuing with Descartes, you might ask why you should believe the supposed evidence is actually true, that dreaming is truly indistinguishable from waking life. The author never says directly what Descartes's reasoning is but instead uses the rhetorical question *Who has not dreamt of "waking up" while still asleep?* to suggest a common human experience. If we made this into an argument, now we would have the following:

- **Evidence**: You have at least once in your life had a dream in which you thought you woke up, only to discover later that you had still been dreaming.
- **Conclusion**: It is impossible to tell the difference between dreaming and being awake.

This may seem peculiar because what was before called the evidence is now the conclusion. Can the exact same statement be both evidence and conclusion? Yes, but only in relation to two different claims. This is nothing special: the same man can be both a son and a father, the same woman both a daughter and a mother. Anytime a claim is used to *support* something else, it's acting as evidence; but whenever it in turn *is supported by* something else, it's acting as a conclusion. If we treated the second argument as a "subargument" within the first, we could represent it like this:

- **Subevidence**: You have at least once in your life had a dream in which you thought you woke up, only to discover later that you had still been dreaming.
- **Subconclusion/evidence**: It is impossible to tell the difference between dreaming and being awake.
- **Conclusion**: Knowledge gained from sense experience is inherently uncertain.

Or, as we'll see more extensively in Chapter 6 of *MCAT CARS Review*, we can use arrows to represent the relevant support relationships, simplifying the depiction of the argument further:

You have at least once in your life had a dream in which you thought you woke up, only to discover later that you had still been dreaming.

It is impossible to tell the difference between dreaming and being awake.

Knowledge gained from sense experience is inherently uncertain.

MCAT Expertise

When authors employ especially difficult arguments, you may find it useful to sketch a diagram on your noteboard next to your Outline, using arrows to denote the support connections. Place the author's main conclusion, or thesis, either at the top with arrows pointing up toward it, or, as we model in the text, at the bottom with down arrows. Multiple pieces of evidence for the same conclusion would appear as branches extending from a common stem; thus, we sometimes call these drawings "argument trees."

When determining relationships such as these, you can use the textual clues discussed in Chapter 3 of *MCAT CARS Review*, especially Evidence and Conclusion keywords (two of the subtypes of Logic keywords). However, as noted there, these keywords are less common than the other two types, and some authors use them quite sparingly. Be alert for hidden support relationships!

COUNTERARGUMENTS

An additional complication to this picture is the **counterargument**, or an argument made against (counter to) a particular conclusion. Counterarguments are also known as *refutations*, *objections*, or *challenges*, and they can be indicated through the use of the aptly named Refutation keywords from Chapter 3 of *MCAT CARS Review*. While some writers will offer *only* counterarguments, their purpose being to argue against some claim, many authors raise counterarguments merely for the sake of refuting them, which is an indirect way to support their conclusions.

The passage from the last chapter provides a key example of this. The author is ultimately interested in arguing in favor of the skeptical position, which can be summarized by the claim *certain knowledge is impossible*. Arguing in favor of a negative is difficult, so the author opts instead to argue *against* Descartes, who tries to support the exact opposite position—that *certain knowledge is possible*. Thus, the author begins the fifth paragraph (playing off a conclusion supported in the previous paragraph) asking, *If even arithmetic can be cast into doubt, then how could anything be known for sure?* The author would just as soon answer this question by saying that nothing *can* be known for sure, but instead gives us Descartes's opposing view. Up to this point, we could summarize the counterargument like so, this time putting the conclusion first:

- **Conclusion**: Certain knowledge is impossible.
- **Refutation**: As long as I think that I am something, I must exist.

If I can be certain that I exist, then certain knowledge *would* be possible, which is *inconsistent* with the conclusion. However, the author is not content to let Descartes win—and yes, for many authors of passages used in CARS, it is effectively about who wins or loses the argument. In the final paragraph, the author responds to the refutation, challenging Descartes's counterargument, thereby supporting the original conclusion. That looks something like this:

- **Conclusion**: Certain knowledge is impossible.
- **Refutation**: As long as I think that I am something, I must exist.
- **Counter-refutation**: Just because there is thinking does not mean there is an "I" who thinks.

Again, keep in mind that refutations will not show up in every CARS passage and that arguments only minimally require two claims: one conclusion and one statement of evidence.

INFERENCES

While an argument could contain as little as one stated conclusion and one piece of stated evidence, most arguments contain hidden or implicit parts as well. The most commonly appearing terms for these are *inferences*, *assumptions*, and *implications*. As verbs, *imply* designates the action of a writer hinting at something without saying it outright, while *infer* indicates the reader's role in deciphering the hints. Whether the unstated conclusion that is reached is called an inference or an implication is simply a difference in emphasis, not meaning. That said, for the sake of precision, whenever we use the word **implication**, we specifically will refer to an unstated conclusion, and **assumption** will be used only for unstated evidence. **Inference** will be used generally to cover any unstated part of an argument, whether an implication or an assumption.

It's important to note that assumptions and implications are not simply claims that are *possibly* or *probably* true given what is said. Rather, inferences must be true or—at the very least—must be highly probable, the most likely option among the alternatives. One way to recognize an inference is by the negative effect it would have on the argument if it were denied. This is a procedure that is employed in the strategy we call the Denial Test, explained in Chapter 10 of *MCAT CARS Review* in the strategy discussion for the appropriately named Inference question type.

In the fifth paragraph of the passage from before, the author actually unearths an assumption in Descartes's argument for the certainty of self-existence by pointing out that *this "I" that exists for certain does not include the physical body, which may just be an illusion, but is simply the thinking self or mind*. We could summarize it as follows:

- **Conclusion**: As long as I think I am something, I must exist.
- **Assumption**: "I" refers only to the mind that thinks and does not include the body.

If this assumption were denied, it would mean that the existence of the physical body could be certain. But can we actually be certain about the existence of our bodies? Descartes himself gives us reason not to think so because he argues earlier that waking and dreaming cannot be distinguished, meaning that I could just be dreaming that I have a body when I'm actually just some brain floating about in a dark laboratory vat. Thus, because Descartes suggests the existence of the body is uncertain, but argues that the existence of the self ("I") is certain, then it must be the case that his concept of the self excludes the body—and this is the assumption that the author identifies.

> **Key Concept**
>
> Unstated claims in arguments are known as inferences. Inferences are either assumptions (unstated evidence) or implications (unstated conclusions).

5: Dissecting Arguments

STRENGTHENING AND WEAKENING

Recall that only claims have truth value; it is a misstatement to say that an argument is true or false. Rather than a simple either/or, the **validity** or **strength** of arguments varies along a continuum with certain proof at one end, complete refutation at the other, and most arguments (virtually all of the ones you see on the MCAT) falling somewhere short of either extreme, with evidence making conclusions more probable and refutations making them less probable.

Questions, particularly both types of Strengthen–Weaken questions (discussed in Chapters 10 and 11 of *MCAT CARS Review*) will often not take the force of a given argument for granted, requiring you instead to imagine how it might be modified. A higher quantity or quality of supporting evidence will *strengthen*, *bolster*, or *further* an argument. This evidence can come in many forms; different fields of study can diverge widely in the types of support that practitioners of the discipline find compelling. For instance, quotations from a novel carry a lot of weight in a literary discussion advocating for an interpretation of that specific piece of fiction, but they would be of much more questionable worth if offered in support of historical claims about events that occurred during its writing.

On the other hand, a claim is said to *weaken*, *challenge*, or *undermine* an argument whenever its truth would make the conclusion more likely to be false. Using the language just introduced, such claims serve as refutations or counterarguments. Note that most refutations do not come in the form of outright contradictions of the original conclusion or evidence. In other words, while inconsistencies will definitely weaken an argument, there are many other ways to pose challenges than simply asserting the opposite. So if an author is, say, making general claims about US citizens on the basis of results from a psychological survey, his argument would be undermined by evidence that the sample used was demographically unrepresentative.

There are only a few basic ways to strengthen or weaken an argument. To see these, it helps to return to the concept of a *table*, using it now as an analogy, with the tabletop substituting for the conclusion and each piece of support represented by a leg of the table. Now, a table might be supported with one sturdy central leg, with two equally balanced legs at either end, or with any number of smaller legs that distribute its weight evenly among them. Regardless of how many legs the table actually has, there are really only three ways to strengthen it (to enhance the support of the tabletop): to add a new leg, reinforce an existing leg, or remove a heavy weight that threatens to bring the table crashing down. Similarly, an argument can be supported by bringing in an entirely new piece of evidence, by providing additional support for an existing piece of evidence, or by challenging a counterargument to its conclusion.

> **Key Concept**
>
> Arguments are evaluated on the basis of their strength or validity, which varies along a spectrum. An argument is strengthened whenever the truth of its conclusion is made more probable and weakened whenever it is made less probable.

In terms of weakening, the same analogy comes to bear. A tabletop will come crashing to the ground if a heavy enough weight is added, just as an argument will collapse if a new refutation is sufficiently weighty. Similarly, supporting a refutation that was raised in the passage (which could be a matter of undermining the author's response to that refutation) is like piling on additional weight. The third and final option for weakening is to knock the legs out from under the table, typically by attacking one that is crucially placed. Analogously, challenging a key piece of evidence can be just as effective at weakening an argument as challenging the conclusion directly.

Conclusion

While academic texts can be vague and subject to interpretation, every question you face on the CARS section will have only one defensible answer. The value of logic is precisely in its ability to clarify your thinking, allowing you to mirror the cognitive processes the testmakers use, and home in on correct responses. As noted in Chapter 1 of *MCAT CARS Review*, of all CARS questions, 30 percent involve *Reasoning Within the Text* and 40 percent require *Reasoning Beyond the Text*. Because even *Foundations of Comprehension* questions can involve making inferences, you can expect more than three-quarters of your score in CARS to benefit from clear logical thinking. To develop your capacity for this even further, we explore two additional applications of logic in the following chapter.

CONCEPT AND STRATEGY SUMMARY

Domains of Discourse
- The **natural domain** corresponds to objects, events, and experiences—everything that can be found in the world around us.
- The **textual domain** corresponds to words, sentences, and paragraphs—everything that directly faces you in an MCAT passage.
- The **conceptual domain** corresponds to concepts, claims, and arguments—everything that underlies logic.

Concepts: The Basic Elements of Logic
- **Concepts** are ideas that have meanings (definitions or connotations).
- Ideas can be related through various relations, which are often indicated with Relation keywords in a passage.
    - Similarity and Difference are very common Relation keywords.
    - Opposition, Sequence, and Comparison keywords are less common, but they are significant when they occur in a passage.
    - The relation between the whole and its parts can be important in some CARS passages.

Claims: The Bearers of Truth Value
- **Claims** can also be called assertions, statements, propositions, beliefs, or contentions.
- Claims are made up of combinations of concepts and relationships of ideas.
- They possess truth value and can thus be true or false.
- Claims can also be related through various relationships.
    - If two claims are **consistent** (compatible or in agreement) with one another, then both can be true simultaneously.
    - If two claims are **inconsistent** (contradictory or conflicting) with one another, then it is impossible for both to be true simultaneously.
    - If one claim **supports** another, then this claim being true would make the other claim more likely to be true as well.
    - If one claim **challenges** (refutes or objects to) another, then this claim being true would make the other claim more likely to be false.

Arguments: Conclusions and Evidence
- At a minimum, arguments contain three parts: a **conclusion**, its **evidence**, and the one-way path of **support** between them.

- **Counterarguments**, also called refutations, objections, or challenges, are the opposite of evidence because they go against the conclusion.
- **Inferences** are unstated parts of arguments.
  - **Assumptions** are unstated pieces of evidence.
  - **Implications** are unstated conclusions.
  - Inferences are claims that must be true or—at the very least—must be highly probable.
- There are three main ways of strengthening an argument:
  - One could provide a new piece of evidence that supports the conclusion.
  - One could support evidence that already exists to support the conclusion.
  - One could challenge refutations against the conclusion.
- There are three main ways of weakening an argument:
  - One could provide a new refutation that goes against the conclusion.
  - One could support refutations that already exist.
  - One could challenge evidence for the conclusion.

Formal Logic

6

6: Formal Logic

In This Chapter

6.1 The Logic of Conditionals   114
     Representing Conditionals   114
     Necessary *vs.* Sufficient
       Conditions   116
     Forming the Contrapositive   116

6.2 Applications of Conditionals   117
     One Form, Many Functions   117
     Justification *vs.* Causation   118
     Logical Translations   119

6.3 Analogical Reasoning   120
     Similarity, Revisited   120
     Arguments from
       Analogy   120
     Extrapolation and
       Integration   122

Concept and Strategy Summary   123

Introduction

LEARNING GOALS

After Chapter 6, you will be able to:

- Recognize conditional relationships between terms
- Differentiate between necessary and sufficient conditions
- Distinguish justification and causation within an argument
- Apply analogical reasoning principles in *Reasoning Beyond the Text* questions

If you think you now know everything that is important to know about logic, then you're in store for a surprise in this chapter. Claims such as the one in the preceding sentence are known as conditionals, and they play a major role in logic. Our present discussion builds on the logical foundations laid in Chapter 5 of *MCAT CARS Review*. We will use the definitions presented in that chapter as a launching point to explore some more formalized aspects of logic. These are the most abstract topics considered in the entire book, so it's essential to be well versed in the established terminology and to proceed slowly and methodically so that the material will not be made even more difficult than it intrinsically is.

Now that you understand the basic anatomy of arguments, we can investigate two special forms of argumentation commonly employed in CARS passages and asked about in questions. The first of these, arguments made using conditional claims, is the more fundamental of the two, with conditional relationships featured in some form in every passage and playing some role in most CARS questions.

Bridge

Your understanding of this chapter is contingent on having internalized the terminology defined in Chapter 5 of *MCAT CARS Review*. Make sure you know the difference between the natural, textual, and conceptual domains, and make sure you can discriminate between a concept, a claim, and an argument.

MCAT Critical Analysis and Reasoning Skills

Consequently, the first part of our discussion explains conditional arguments' basic structure and operation, while the next part examines how conditionals manifest themselves in CARS passages and questions, especially in the commonly appearing relationships of justification and causation. The final part of the chapter is devoted to the other major form of argument, analogical reasoning, employed by many authors of passages used in CARS and tested in almost every question type in the category the AAMC calls *Reasoning Beyond the Text*.

6.1 The Logic of Conditionals

The term **conditional** ultimately refers to not just one but an entire class of unidirectional relationships between two items. These items could be from any of the domains discussed in the previous chapter (the natural, textual, or logical), but we'll use a word from the textual domain, **terms**, to discuss the items generically. Different kinds of conditionals will be discussed in the second part of the chapter, but all of them have the same basic form and logical characteristics.

REPRESENTING CONDITIONALS

Assuming that **X** and **Y** are two terms, and that the arrow ($\rightarrow$) indicates the direction of the connection, the following formulation expresses the conditional relationship:

$$X \rightarrow Y$$

Consider this a representation of the basic concept of the conditional. But what does this really mean? What exactly is the connection signified by the arrow? Let's assume for a moment that the terms **X** and **Y** denote claims, which (as detailed in Chapter 5 of *MCAT CARS Review*) are evaluated on the basis of their truth value—the potential to be either true or false (but not both at the same time). When working with claims, we could rewrite the conditional relationship in this slightly less formal way:

If **X** is true, then **Y** is true.

Note that while it is made up of a relationship between two claims, this is itself a kind of statement with its own truth value. We'll call this a **conditional claim**; that is, it is a statement that has a meaning that could be captured equivalently by using an **if–then** assertion, as above. Moreover, adopting the common convention, we'll designate the *if* term the **antecedent** and the *then* term the **consequent**. Here's a concrete example, with the antecedent and consequent in bold:

If it's true that **you are in Pennsylvania**, then it's also true that **you are in the United States of America**.

> **Key Concept**
> The conditional relationship is formally defined as the impossibility of having a true antecedent and a false consequent simultaneously. Technically speaking, if the antecedent and consequent terms are items other than claims (such as causally connected events, considered below), *true* and *false* would be replaced by analogous evaluations (so, for cause and effect, the terms *present* and *absent* might be substituted).

Or more simply:

> If **you're in Pennsylvania**, then **you're in the United States.**

What is distinctive about such claims is that their truth, formally speaking, depends on only the truth of the antecedent and consequent. As long as it's impossible for **X** to be true at the same time that **Y** is false (that you could be in Pennsylvania without being in the United States), then the conditional statement is true. In other words, the **conditional relationship** can be defined as the impossibility of having a true antecedent and a false consequent.

This can be represented formally in what is known as a **truth table**. In Table 6.1, all four possible combinations of the truth of **X** and **Y** are presented, as well as the resultant truth of the conditional claim in the final column:

X	Y	X → Y
true	true	**true**
true	false	**false**
false	true	**true**
false	false	**true**

Table 6.1. Truth Table for Conditional Claims

Returning to the location example, we know that it's simply not possible for the antecedent (*you're in Pennsylvania*) to be true while the consequent (*you're in the United States*) is false. This is because Pennsylvania is completely contained within the United States. Thus, we can say with confidence that the statement *If you're in Pennsylvania, then you're in the United States of America* is true.

Notice that this does not work the other way around; a conditional is unidirectional. You say something entirely different, something that is factually false, when you claim, *If you are in the United States, then you are in Pennsylvania*. In addition to the other 49 states, the District of Columbia, and 16 territories including Puerto Rico, Guam, and American Samoa, the United States occupies many military bases and diplomatic embassies, effectively spanning the globe. As can be seen, there are many ways to be in the United States without being in Pennsylvania.

Now here is a crucial point that cannot be stressed enough: *a conditional claim can still be true even if you know the antecedent is itself false*. In our geographical example, it doesn't matter whether you're in Maryland (in which case **X** is false but **Y** is true) or in New Zealand (for which both **X** and **Y** would be false). Neither of those is what might be thought of as the forbidden combination: a true antecedent plus a false consequent. No matter how hard you try, there's simply no way to be in

Pennsylvania without also being in the United States (that's true even for the foreign embassies and consulates located in Philadelphia, which would count as being on neither US nor Pennsylvania soil, from the standpoint of international law).

NECESSARY vs. SUFFICIENT CONDITIONS

Besides using the language of *antecedent* and *consequent*, there is another noteworthy way of understanding the relationship embodied in a conditional that divides the relationship in two by considering it from the perspective of each term. In this alternative account, each of the two terms in $X \rightarrow Y$ counts as a type of condition. Thus, **X** would be the **sufficient condition**, while **Y** would be the **necessary condition**. Or, to state it in a way that highlights the relation between the two terms, in $X \rightarrow Y$, **X** is a sufficient condition of **Y**, while **Y** is a necessary condition of **X**. Let's consider each side of this, in turn.

> **Key Concept**
>
> Another way of conceptualizing a conditional is to think of it as two simultaneous relationships. In the conditional claim *if* **X**, *then* **Y**, **X** is a sufficient condition of **Y**, while **Y** is a necessary condition of **X**.

Saying that **X** is **sufficient** for **Y** is equivalent to saying that it's impossible to have **X** without also having **Y**. So, to return to our intuitive, geographical example, it's enough to know that you're in Pennsylvania—it doesn't matter whether it's Pittsburgh, Philadelphia, or Punxsutawney—to be able to conclude that you're in the United States. In short, being in Pennsylvania is *sufficient* for being in the United States. This is really just another way of expressing the conditional relationship as the impossibility of a true antecedent and a false consequent.

> **MCAT Expertise**
>
> In more complex cases, you'll find that there can be more than one necessary condition or more than one sufficient condition. Suppose that both **X** and **Y** together lead to **Z**. We would then say that **X** and **Y** are *mutually necessary and jointly sufficient* for **Z**. For instance, to be a mother it is not simply sufficient to be female, but one also needs to be a parent. Thus, being female (**X**) and being a parent (**Y**) are mutually necessary and jointly sufficient for being a mother (**Z**). Though rare, this complication has appeared in past MCAT exams, including once in an extremely challenging question on linguistics that required understanding the two notions, only briefly explained in the accompanying passage.

While the relationship of sufficiency is the aspect of the conditional that we're already familiar with, necessity is a bit new, although it's simply the same relationship considered from the opposite perspective. Saying that **Y** is **necessary** for **X** is like saying that **Y** is a **prerequisite** or **requirement** of **X**. This reversal amounts to a negative claim: if **Y** is not true, then **X** is not true. If we consider the example from before from this perspective, it becomes:

If you are **not in the United States**, then you are **not in Pennsylvania**.

This statement is logically equivalent to *If you are in Pennsylvania, then you are in the United States*. However, even though the two versions of the claim will always have the same truth value, their meanings are not identical—if nothing else, there is a difference in point of reference. This alternative form, which emphasizes a necessary condition, is even known by a special name, the contrapositive.

FORMING THE CONTRAPOSITIVE

Whenever an author makes any kind of conditional claim, it is always possible to translate the conditional claim into other relationships. Most notably, one can form the logically equivalent assertion known as the contrapositive. By definition, the

6: Formal Logic

contrapositive of *if X, then Y* is *if not Y, then not X*. Alternatively, this can be represented using a tilde (∼) to stand for negation (∼**X** thus means *not X* or *the negation of X*):

$$\text{Conditional: } X \rightarrow Y$$

$$\text{Contrapositive: } \sim Y \rightarrow \sim X$$

One of the most useful reasons for forming the contrapositive is that it's a guaranteed inference, a logical equivalent for any conditional claim made in a passage. Whenever you recognize Logic keywords or one of the English translations of a conditional claim, as listed in Table 6.2, you'll want to keep in mind that a rendition of its contrapositive could appear as a correct answer.

> **Key Concept**
>
> The contrapositive is a statement with a different connotation (emphasizing the necessary condition, rather than the sufficient) but that is logically equivalent to a conditional assertion. For instance, the contrapositive of *if X is true, then Y is true* can be written as either *if Y is not true, then X is not true* or *if Y is false, then X is false*.

If **X**, then **Y**.	If I'm in Pennsylvania, then I'm in the United States.
X is sufficient for **Y**.	Being in Pennsylvania is sufficient for being in the United States.
All **X** are **Y**.	All those in Pennsylvania are in the United States.
X only if **Y**.	I'm in Pennsylvania only if I'm in the United States.
If not **Y**, then not **X**. [Contrapositive]	If I'm not in the United States, then I'm not in Pennsylvania.
Y is necessary for **X**.	Being in the United States is necessary for being in Pennsylvania.
Only **Y** are **X**.	Only those in the United States are in Pennsylvania.
Not **X** unless **Y**.	I'm not in Pennsylvania unless I'm in the United States.

Table 6.2. English Translations of X → Y

6.2 Applications of Conditionals

Now that you've seen how conditionals work as logical structures, let's consider how they function in CARS passages and questions. Remember that with each of the following cases, although the logical form of the relationship is identical, the actual relationships are not equivalent because they operate within the different domains of discourse described in Chapter 5 of *MCAT CARS Review*.

ONE FORM, MANY FUNCTIONS

Although we have been considering only claims as the antecedent and consequent terms in a conditional, these terms could take some other form, as noted earlier. For example, the consequent could be a concept that is an essential part of the definition of a larger concept, which would be the antecedent. This would be an example of a whole–parts relationship, which you can now see can be represented as a conditional in which the *whole* being defined is an antecedent or sufficient condition and any defining *part* is a consequent or necessary condition.

For instance, consider the relationship between the concept *unicorn* and the concept *horn*, from an example first posed in the previous chapter: unicorn → horn. The first idea is a sufficient condition for the second: if you know you're dealing with a unicorn, then you know you're dealing with a concept that includes a horn in its definition. Simultaneously, the consequent is a necessary condition of the antecedent (forming the contrapositive): no horn entails no unicorn. Like any conditional, this doesn't work the other way around: a horn doesn't guarantee a unicorn because the concept of a narwhal, a rhinoceros, or even certain lizards would also include a horn in its definition.

Other instances of conditionals will fall under the natural domain of discourse rather than the conceptual. For example, an object and its essential components or characteristics (more whole–parts relationships) could be represented as conditionals: *object → components* or *object → characteristics*. A case of the first is the connection between a living human body and an intact brainstem—though other parts may be present, no human can live independently without a functioning brainstem. The latter can be seen in the relationship between plant cells and cell walls. While other organisms can also have cell walls (fungi, for example, have a cell wall composed of *chitin*), all known plant cells possess a cell wall. Thus, being a plant cell is sufficient for having the characteristic of a cell wall.

These are but a few of the possible forms that a conditional relationship could take. Yet while there are many specific iterations, two in particular stand out as being by far the most common. Moreover, as originally suggested in Chapter 3 of *MCAT CARS Review*, these two relationships are the ones most commonly revealed with the use of Logic keywords.

JUSTIFICATION *vs.* CAUSATION

The term **justification** is ultimately another way of talking about the relationship of logical support. In this particular relationship, the two items in the conditional are claims, aspects of the conceptual domain of discourse: the antecedent is the **evidence**, and the consequent is the **conclusion**. In fact, any argument can be made into a single conditional relationship if all the evidence is lumped together into one single statement with a lot of conjunctions (*and*s).

While justification is a kind of conceptual conditional connecting two claims, **causation** is another natural connection, one that links two *events*. Unlike a mere **correlation**, in which two occurrences are found to accompany one another, causation is a one-way relationship, like any conditional. In this instance, however, the order is temporal: the antecedent (cause) must occur before the consequent (effect). Moreover, in contrast to sequences of events that may happen coincidentally, causes are directly responsible for leading to their effects. Proving this last characteristic of causality can be particularly challenging, so robustly supporting a causal assertion is no easy task.

Both justification and causation are conditional relationships, but they are quite distinct. On the one hand, causes and effects are *events* in the *natural* world, with the cause always preceding the effect in time. On the other hand, pieces of evidence and conclusions are *claims* about the world, part of the conceptual domain of discourse, with a relationship that might not follow the arrow of time. This is evident in the following contrast:

The dinosaurs were turned into fossils **because of** an asteroid's catastrophic impact.

vs.

We know that there was a catastrophic asteroid impact **because of** evidence provided by dinosaur fossils.

The second argument is distinct insofar as it does not address the dinosaurs and the asteroid directly, as the first one does, but rather concerns our *knowledge* of the asteroid and the *evidence* that the fossils provide. While the asteroid may be (at least indirectly) the *cause* of the fossils, without a time machine we have access only to what remains of the fossils in the present day. Because we are able to see only the *effect*, we have to make an inference to figure out the cause.

The upshot of this is that whenever you encounter Evidence or Conclusion keywords, you should always clarify the domain of discourse in which the author is writing. If the terms in the conditional are natural, then *because* indicates a cause, and *consequently* denotes an effect. When the terms are conceptual entities instead (specifically claims with truth value), words like *since* point to evidence, while *therefore* and *thus* designate conclusions. Your default assumption should be justification because this relationship tends to be more commonly tested.

> **Bridge**
> Hill's criteria can be used to support the likelihood of causality in a relationship. While only temporality is absolutely necessary for causation, it is not sufficient. The other criteria include strength, dose–response relationship, consistency, plausibility, consideration of alternate explanations, experimentation, specificity, and coherence. Hill's criteria are discussed in Chapter 11 of *MCAT Physics and Math Review*.

LOGICAL TRANSLATIONS

While as a general rule, an Evidence keyword precedes an antecedent or sufficient condition and a Conclusion keyword precedes a consequent or necessary condition, there are some ways of expressing a conditional relationship that are less clear about distinguishing the two terms. To aid you in recognizing some of these more elaborate articulations, note that each of the terms in Table 6.2 is logically equivalent to the basic conditional statement **X→Y** (*If I'm in Pennsylvania, then I'm in the United States*).

MCAT Critical Analysis and Reasoning Skills

6.3 Analogical Reasoning

The final formal application of logic worth studying for Test Day is associated almost exclusively with one category of questions. That said, as you'll see in Chapter 11 of *MCAT CARS Review*, the *Reasoning Beyond the Text* category constitutes two-fifths of all CARS questions. These are relatively easy to identify because they always involve new information (in the question stem, the set of answer choices, or both) that is not stated or even suggested by the passage and that may not even seem to be related at first. In one typical pattern, words like *Suppose*, *Assume*, and *Imagine* precede an elaborate scenario that fills several lines of text, followed up by a question that connects this new content to the author or passage. Making the appropriate connections to address such questions requires a skill that may seem so obvious that it defies explanation: recognizing similarities.

SIMILARITY, REVISITED

In Chapter 3 of *MCAT CARS Review*, we noted how Similarity keywords reveal continuities within texts, often cluing you into repetitions or ancillary details. These tend to be less significant than contrasts and other more complex relationships, signaled by the other Relation keywords. Nevertheless, similarity as a concept is actually far more important than this strategic guideline would initially suggest.

Similarity is one of those ideas that seem obvious until you try to explain them. At its most extreme, similarity becomes *identity*, as would be the case when two terms are used to reference one single natural entity: the so-called *Morning Star* and *Evening Star* are identical because both terms refer to the planet Venus. Slightly less extreme cases of similarity would be exact duplicates, such as photocopies of documents, backups of computer files, the genetics of monozygotic twins, coins of the same denomination, or interchangeable machine parts.

With CARS passages and questions, similarity is typically more than a simple matter of two terms that refer to the same thing or duplicates. Most concepts that you'll be comparing will be complex, with many characteristics, some of which will overlap and some of which will not. This complexity depends upon the relationship between a whole and its parts, introduced in the preceding chapter and noted earlier as a form of conditional. These more involved comparisons, known as **analogies**, have special considerations governing their use in arguments.

ARGUMENTS FROM ANALOGY

An **argument from analogy** begins on the basis of similarities between two things to argue for an additional commonality between them. In other words, it concludes that something (typically an *object* or *event* in the natural domain, though potentially

Bridge

You may not have realized it as you were studying Chapter 2 of *MCAT Organic Chemistry Review*, but isomerism offers a perfect illustration of the complexities within the concept of similarity. The key to understanding isomerism is to pay attention to the particular similarities and differences that make each isomeric relationship precisely what it is. For example, conformational isomers, which differ only in rotation around a σ bond, are much more alike than structural (constitutional) isomers in which the bonds are not even identical, which is why conformers tend to share so many more properties. On the CARS section, you'll be relating concepts rather than chemicals, but you'll still want to break down the specifics. *Reasoning Beyond the Text* questions become far more manageable if you can isolate the exact points of likeness that new ideas and claims share with passage information, as well as those places in which the two diverge.

6: Formal Logic

a complex *concept* instead) has a certain quality because it is like something else with a similar characteristic. Analogies always contain two terms: the **known** entity is the one with characteristics that are already well established, while the **unknown** entity is only partially understood. If we call the known **K** and the unknown **U**, we arrive at the following formalization of an argument from analogy:

- **Evidence: K** and **U** share similar corresponding characteristics (K_1 and U_1, K_2 and U_2, K_3 and U_3, and so on).
- **Evidence: K** possesses characteristic *Kx*.
- **Conclusion: U** will possess similar characteristic *Ux*.

This type of reasoning is employed regularly in biological contexts. For instance, Charles Darwin's explanation of the origin of new species is based on an analogy from agriculture and animal husbandry. If *artificial selection* can account for dramatic differences in the appearance of domesticated plants and animals over just a few generations, Darwin reasoned, then perhaps *natural selection* could account for speciation over longer stretches of time. Mapping onto the formal version above, **K** (the known) would be the crops and livestock found on farms, while **U** (the unknown) would be all the species in nature, and the *x* terms would concern the selectors that cause species divergence.

The strength of arguments from analogy can vary widely. An analogy will be weak if it depends on only one minor point of similarity between the known and the unknown or if it draws upon commonalities (regardless of quantity) that are irrelevant to the conclusion. Just like in genetics, in which a mutation to a noncoding region of the genome tends to have no consequences for the organism's phenotype, a point of contrast between a new scenario and the passage could be insignificant to the question at hand. For example, it would be a mistake to infer, as did many ancient cultures, that the sun and the moon are living creatures because they move across the sky like birds. Though celestial bodies and avian fauna share this superficial feature, the reasons behind their motion are vastly divergent, so the analogy does not hold.

Although analogies are often located within passages themselves (making them fair game for questions that would fall under the other categories), analogical reasoning is most often tested in CARS through the use of *Reasoning Beyond the Text* questions. Your task with these will be to isolate the known from the unknown and assess their corresponding points of analogy. There is no easy formula for judging how similar two situations are or for determining how relevant a point of commonality is for an analogy, but these are skills that will improve with additional CARS practice. As a general rule, the MCAT tends to ask about "deeper" points of analogy, going beneath surface characteristics to underlying structural similarities. So, if a question

Key Concept

An argument from analogy is a kind of justification in which the two pieces of evidence are the similarity between two terms and the known existence of a particular characteristic in one of the two. The conclusion is that the unknown term will also possess the same or a similar characteristic.

Bridge

Reasoning Beyond the Text questions, described in Chapter 11 of *MCAT CARS Review*, make up 40 percent of the questions you'll see on Test Day. These questions share one common feature: asking for the connections between the passage and new information. Oftentimes, these questions focus on analogical arguments.

MCAT Critical Analysis and Reasoning Skills

asks for an item analogous to a forged painting, the correct answer would more likely involve deception (such as a disguise or a lie) than artistic representation (such as a photograph or sculpture).

EXTRAPOLATION AND INTEGRATION

There are two basic ways in which analogical reasoning is employed in CARS. In one case, the passage provides the known term, with its various characteristics, and the question gives the new context that establishes the unknown. In such questions, you are being asked to apply, extend, or extrapolate the ideas from the passage to a new situation, so you can take the information from the passage as a given. Kaplan thus calls these Apply questions, as described in Chapter 11 of *MCAT CARS Review*.

In the other case, you will have to integrate or incorporate new ideas into the passage, making a judgment about the effect that they would have. For these questions, the information from the text is subject to change, and so we treat it as though it were the unknown, while the question stem typically establishes the known entity. The key to answering these effectively will be understanding how the new information provided is analogous to evidence, refutations, or conclusions from the passage itself. Because this question type usually deals with strengthening and weakening arguments, Kaplan calls these Strengthen–Weaken (Beyond the Passage) questions, which, as you'll see, are closely tied to the Strengthen–Weaken (Within the Passage) type, featured in Chapter 10 of *MCAT CARS Review*.

Conclusion

So ends the discussion of a few useful formal applications of logic. Appreciating the general characteristics of conditionals and analogies, as well as the ways in which each argumentative form manifests itself in passages and questions, will clarify your thinking and garner you many additional points. Given your new grasp of logic, we can now turn to consideration of the passages one last time and to the additional depths of understanding that can be gleaned when we integrate rhetorical analysis from Chapter 2 of *MCAT CARS Review* with not only the keywords and Outline strategies from Chapters 3 and 4 but also with the tools of logic just showcased in Chapters 5 and 6.

CONCEPT AND STRATEGY SUMMARY

The Logic of Conditionals

- A **conditional** is a unidirectional relationship that exists between two terms.
  - Conditionals can be represented with language (*if X, then Y*), or symbols: $X \to Y$.
  - The **antecedent** (**X**) can also be called a **sufficient condition**, **evidence** (in cases of justification), or **cause** (in cases of causation).
  - The **consequent** (**Y**) can also be called a **necessary condition**, **conclusion** (in cases of justification), or **effect** (in cases of causation).
- A **conditional claim** is true if it is impossible to have a true antecedent and a false consequent simultaneously.
- Operations of formal logic can be represented in a **truth table**.
- **Sufficiency** refers to the impossibility of having an antecedent without its consequent.
- **Necessity** refers to the idea that if the consequent is not true, then the antecedent is also not true.
  - Necessity can be illustrated as *if not Y, then not X* or $\sim Y \to \sim X$.
  - This conditional can also be called the **contrapositive**.
  - The contrapositive is logically equivalent to the original conditional, but it carries a different connotation.

Applications of Conditionals

- One common application of conditionals is the **whole–parts relationship**.
  - One concept can be a part of another concept (the whole) in the conceptual domain.
  - One component or characteristic can be part of an object in the natural domain.
- **Justification** is the relationship of logical support between a piece of evidence and its conclusion.
- **Correlation** is the relationship of two events accompanying one another.
- **Causation** is the one-way relationship of the antecedent leading to the consequent (cause and effect).
- There are many ways of writing the same logical relationship in English.

Analogical Reasoning

- **Argument from analogy** uses the similarities between two things to argue for an additional commonality between them.
  - The **known** entity is the one with characteristics that have already been established.
  - The **unknown** entity is the one that is only partially understood.
- The evidence in an analogical argument is twofold, leading to a single conclusion:
  - One piece of evidence is that the known and unknown entities share similar corresponding characteristics.
  - The other evidence is that the known entity possesses some characteristic of interest.
  - The conclusion drawn from this evidence is that the unknown entity also possesses a similar, relevant characteristic.
- An analogy can be strengthened by greater similarity between the known and unknown.
  - The more points of similarity between the two, the stronger the analogy
  - The more relevant (structural as opposed to superficial) the similarities between the two, the stronger the analogy
  - The fewer relevant differences between the two, the stronger the analogy

7

Understanding Passages

7: Understanding Passages

In This Chapter

7.1 Varieties of Passages   **128**
    Humanities   129
    Social Sciences   131

7.2 Support in Passages   **133**
    Categories of Support   134

7.3 Anticipating Questions   **136**

Concept and Strategy Summary   **142**

Worked Example   **144**

Introduction

> **LEARNING GOALS**
>
> After Chapter 7, you will be able to:
>
> - Categorize MCAT passages as Scientific, Historical, Arts, or Philosophical passages
> - Recognize support for a claim within an MCAT passage
> - Given an MCAT passage, predict questions that are likely to be asked about it

This chapter is a turning point in *MCAT Critical Analysis and Reasoning Skills Review*. Whereas the previous chapters focused on insights into rhetoric, the modes of reading, and keywords; the Kaplan Method for CARS Passages; and arguments and logic; the chapters that follow will focus on the Kaplan Method for CARS Questions and question types. This is the same transition you'll make 39 times on Test Day as you shift from critically reading a passage to answering its questions. All of the strategies we've discussed thus far come together in this chapter as we look at how the different passage elements can work together to improve your understanding of the passages you encounter in the *Critical Analysis and Reasoning Skills* (CARS) section. All the while, we are looking ahead to the real source of points: correctly answering the questions.

This chapter is broken down as follows. First, we'll examine some characteristic varieties of passages that you can expect to see on Test Day while considering a few common themes that exist in each variety. Second, we'll discuss the different kinds of support that are used in passages and how these kinds of support relate to the variety of passages introduced. Both the passage variety and the types of support that are used feed into the same common goal, which we address at the end of this chapter:

MCAT Critical Analysis and Reasoning Skills

anticipating the questions that will be asked for a given passage. Anticipating the questions profoundly improves your speed and accuracy in the CARS section—it's much quicker and easier to answer a question you've been waiting for all along!

7.1 Varieties of Passages

While the AAMC lists eleven different fields in the humanities and a dozen in the social sciences, as shown in Table 7.1, most CARS passages can be classified into one of a handful of categories. While it's true that each discipline has its own distinctive characteristics, similar fields share certain tendencies. Being aware of these similarities can help shape your expectations about what a passage will include as you Scan it in the first step of the Kaplan Method for CARS Passages. Moreover, as we'll see when we return to these examples in the final section of this chapter, you can use these common varieties to help predict the types of questions likely to be asked.

MCAT Expertise

According to the AAMC, 50 percent of the questions in the CARS section will be from the humanities, and 50 percent will be from the social sciences.

Humanities	Social Sciences
Architecture	Anthropology
Art	Archaeology
Dance	Economics
Ethics	Education
Literature	Geography
Music	History
Philosophy	Linguistics
Popular Culture	Political Science
Religion	Population Health
Studies of Diverse Cultures*	Psychology
Theater	Sociology
	Studies of Diverse Cultures*

* Note: Studies of Diverse Cultures can be tested in both humanities and social sciences passages.

Table 7.1. Humanities and Social Sciences Disciplines in the CARS Section[1]

In order to offer you a better sense of the kind of variation that passages can bring, but without overwhelming you with the full 500–600 words of actual CARS passages, we introduce in this chapter a series of four mini-passages, each of which is about half the length of an ordinary CARS passage. We'll look at a number of characteristics of passages in this chapter, using them as illustrations. In Chapters 9 through 11 of *MCAT CARS Review*, which analyze the various question types, we'll return to these passages with sample questions that mimic what you could expect to see on Test Day. These

---

1. AAMC, *The Official Guide to the MCAT 2015 Exam* (Washington, D.C.: Association of American Medical Colleges, 2014), 311–22.

examples are certainly not exhaustive; many passages will blend characteristics of the different varieties, and some passages will not cleanly fit into any of them. The benefit of determining the variety of the passage is to enable you to do something you may do unconsciously in the science sections: set your expectations appropriately. In a passage on rate kinetics, you expect to see rate laws, rate-of-formation tables, and reaction mechanisms. You anticipate that the questions will ask you to determine a rate law using the tables and to consider topics like rate-limiting steps and catalysis. Similarly, you can set expectations about the trajectory of a CARS passage by knowing its variety and can anticipate what questions you'll see on that passage. These expectations allow you to navigate the passage more quickly and to start racking up points with correct answers.

> **Key Concept**
>
> Identifying the variety of the passage will give you insight into the likely form and structure of the passage, as well as the likely questions you'll be asked. This will make you more efficient in navigating the passage and more rapid and accurate in answering the questions.

HUMANITIES

Passages in the humanities tend to fall into two broad categories. The first category, which includes most of the passages from architecture, art, dance, literature, music, popular culture, and theater, could broadly be considered **Arts passages**. They tend to take a form like the following:

An Arts Passage

One of the first examples of the ascendance of abstraction in 20th-century art is the Dada movement, which Lowenthal dubbed "the groundwork to abstract art and sound poetry, a starting point for performance art, a prelude to postmodernism, an influence on pop art…and the movement that laid the foundation for surrealism." Dadaism was ultimately premised on a philosophical rejection of the dominant culture, which is to say the dominating culture of colonialist Europe. Not content with the violent exploitation of other peoples, Europe's ruling factions once again turned inward, reigniting provincial disputes into the conflagration that came to be known by the Eurocentric epithet "World War I"—the European subcontinent apparently being the only part of the world that mattered.

The absurd destructiveness of the Great War was a natural prelude to the creative absurdity of Dada. Is it any wonder that the rejection of reason made manifest by senseless atrocities should lead to the embrace of irrationality and disorder among the West's subaltern artistic communities? Marcel Janco, one of the first Dadaists, cited this rationale: "We had lost confidence in our culture. Everything had to be demolished. We would begin again after the *tabula rasa*." Thus, we find the overturning of what was once considered art: a urinal becomes the *Fountain* after Marcel Duchamp signs it "R. Mutt" in 1917, the nonsense syllables of Hugo Ball and Kurt Schwitters transform into "sound poems," and dancers in cardboard cubist costumes accompanied by foghorns and typewriters metamorphosize into the ballet *Parade*.

Unsurprisingly, many commentators, including founding members, have described Dada as an "anti-art" movement. Notwithstanding such a designation, Dadaism has left a lasting imprint on modern Western art.

Sample Outline

P1. Dada (Author + +): early abstract art, originated in rejection of European culture (Author − −)

P2. WWI led to Dada's "creative absurdity"; example artworks

Goal: To discuss the impact, origins, and characteristics of Dadaism

Notice the heavy use of quotations from both artists and critics, the strong opinions evident in the passage's Author keywords, and the use of descriptive language to illustrate artistic examples. While not every Arts passage will take this form—literature passages, in particular, can show a lot of variation beyond these tendencies—many will.

Most of the other passages in the humanities fall into the second category, which includes ethics, philosophy, religion, and studies of diverse cultures and which might be called **Philosophical passages** more broadly. The following, from the field of ethics, is such a passage:

An Ethics Passage

The most prevalent argument against doctor-assisted suicide relies upon a distinction between *passive* and *active* euthanasia—in essence, the difference between killing someone and letting that person die. On this account, a physician is restricted by her Hippocratic oath to do no harm and thus cannot act in ways that would inflict the ultimate harm, death. In contrast, failing to resuscitate an individual who is dying is permitted because this would be only an instance of refraining from help and not a willful cause of harm. The common objection to this distinction, that it is vague and therefore difficult to apply, does not carry much weight. After all, applying ethical principles of *any sort* to the complexities of the world is an enterprise fraught with imprecision.

Rather, the fundamental problem with the distinction is that it is not an ethically relevant one, readily apparent in the following thought experiment. Imagine a terminally ill patient hooked up to an unusual sort of life support device, one that only functioned to prevent a separate "suicide machine" from administering a lethal injection so long as the doctor pressed a button on it once per day. Would there be any relevant difference between using the suicide machine directly and not using the prevention device? The intention of the doctor would be the same (fulfilling the patient's wish to die), and the effect would be the same (an injection causing the patient's death). The only variance here is the means by which the effect comes about, and this is not an ethical difference but merely a technical one.

Sample Outline

P1. "Most prevalent argument" against euthanasia assumes passive/active difference (vague is okay)

P2. Author: distinction is not ethically relevant (supported w/thought experiment)

Goal: To argue the distinction between passive and active euthanasia is not ethically relevant

Philosophical passages tend to be abstract and heavy on logic—the Descartes passage from Chapter 4 of *MCAT CARS Review* was another such example. They focus heavily on concepts and the relations between them, and they often appeal to the reader's memory or imagination, drawing on common experiences or thought experiments (as in the second paragraph of this passage).

Keep in mind that there will be plenty of humanities passages that mix characteristics of Arts and Philosophical passages, as well as some passages that don't properly fit in either category. Nevertheless, this distinction is useful (as we'll see below) for setting expectations about the kind of support that a passage will use as well as the types of questions that will accompany it.

SOCIAL SCIENCES

When it comes to the social sciences, some passages take what might be called a **Scientific** form, such as passages in anthropology, education, linguistics, population health, psychology, and sociology. Here's an example from psychology:

A Psychology Passage

There is no shortage of evidence for the existence of systemic biases in ordinary human reasoning. For instance, Kahneman and Tversky in their groundbreaking 1974 work proposed the existence of a heuristic—an error-prone shortcut in reasoning—known as "anchoring." In one of their most notable experiments, participants were exposed to the spin of a roulette wheel (specially rigged to land randomly on one of only two possible results) before being asked to guess what percentage of United Nations member states were African. The half of the sample who had the roulette wheel stop at 65 guessed, on average, that 45% of the UN was African, while those with a result of 10 guessed only 25%, demonstrating that prior presentation of a random number otherwise unconnected to a quantitative judgment can still influence that judgment.

> **Bridge**
>
> Does the content of this passage look familiar? Heuristics are simplified principles used to make decisions; they are colloquially called "rules of thumb." Bias is a systematic error made in reasoning. These topics are discussed in Chapter 4 of *MCAT Behavioral Sciences Review*.

MCAT Critical Analysis and Reasoning Skills

The anchoring effect has been observed on repeated other occasions, such as in Dan Ariely's experiment that used digits in Social Security numbers as an anchor for bids at an auction, and in the 1996 study by Wilson *et al.* that showed even awareness of the existence of anchoring bias is insufficient to mitigate its effects. The advertising industry has long been aware of this bias, the rationale for its frequent practice of featuring an "original" price before showing a "sale" price that is invariably reduced. Of course, anchoring is hardly alone among the defective tendencies in human reasoning; other systemic biases have also been experimentally identified, including loss aversion, the availability heuristic, and optimism bias.

Sample Outline

P1. Systemic biases in reasoning: ex. anchoring heuristic (evidence: Kahneman & Tversky)

P2. More evidence for anchoring (Ariely, Wilson *et al.*); other systemic biases

Goal: To present evidence for systemic biases in reasoning, especially anchoring

MCAT Expertise

While psychology and sociology can be tested in both the *Critical Analysis and Reasoning Skills* section and the *Psychological, Social, and Biological Foundations of Behavior* section, the former will not require outside knowledge. In fact, bringing in outside knowledge to answer any question in the CARS section can lead you astray, drawing you toward Out of Scope answer choices.

In this passage, the numbers should jump off the page, along with the heavy reference to empirical studies as support (discussed more below). The author's opinion is less glaring than in an Arts passage, but there are still indications of attitude with Author keywords like *groundbreaking* and *notable* in the first paragraph. In fact, a Scientific passage could come to look more like a passage in the *Psychological, Social, and Biological Foundations of Behavior* section.

The counterparts to Scientific passages are **Historical passages,** which include many examples from archaeology, economics, geography, history, political science, and studies of diverse cultures. Note that this type is far more variable than the others and can sometimes even look a little bit more like a humanities passage, such as in this example:

A History Passage

In 1941, an exuberant nationalist wrote: "We must accept wholeheartedly our duty and our opportunity as the most powerful and vital nation...to exert upon the world the full impact of our influence, for such purposes as we see fit and by such means as we see fit." If forced to guess the identity of the writer, many US citizens would likely suspect a German jingoist advocating for *Lebensraum*. In actuality, the sentiment was expressed by one of America's own: Henry Luce, the highly influential publisher of the magazines *Life*, *Time*, and *Fortune*. Luce sought to dub the 1900s the "American Century," calling upon the nation to pursue global hegemony as it slipped from the grasp of warring Old World empires. As a forecast of world history, Luce's pronouncement seems prescient—but is it justifiable as a normative stance?

Not all of Luce's contemporaries bought into his exceptionalist creed. Only a year later, Henry Wallace, vice president under FDR, insisted that no country had the "right to exploit other nations" and that "military [and] economic imperialism" were invariably immoral. It is a foundational assumption in ethics that the wrongness of an act is independent of the particular identity of the actor—individuals who pay no heed to moral consistency are justly condemned as hypocrites. So why should it be any different for nation-states? In accord with this principle, Wallace proselytized for "the century of the common man," for the furtherance of a "great revolution of the people," and for bringing justice and prosperity to all persons irrespective of accidents of birth. Sadly, Wallace never had the chance to lead the United States in this cosmopolitan direction; prior to Roosevelt's demise at the beginning of his fourth term, the vice presidency was handed to Harry Truman, a man whose narrow provincialism ensconced him firmly in Luce's camp. And with Truman came the ghastly atomic eradication of two Japanese cities, the dangerous precedent set by military action without congressional approval in Korea, and a Cold War with the Soviet Union that brought the world to the brink of nuclear destruction.

Sample Outline

P1. Luce: "American Century" = United States has right to influence world (Author: — — —)

P2. US V.P. Wallace (+ + +) rejected Luce's idea, but replaced by Truman (— — —) who supported it

Goal: To describe Luce and Wallace's opposing views on US power, favoring Wallace

Not all historical passages will be this heavily opinionated, but they will tend to draw on historical events and quotations from sources alive at the time of the events discussed. In passages in history, economics, and political science, you'll occasionally find empirical studies, which make them more like Scientific passages, or heavy theoretical discussions, which make them more Philosophical.

7.2 Support in Passages

Because a majority of questions in the CARS section will have some connection to logical support, it's essential to understand the different kinds of support that can be found in CARS passages.

MCAT Critical Analysis and Reasoning Skills

CATEGORIES OF SUPPORT

Unsupported Claims

Authors say a lot of different things in CARS passages—a single long paragraph could potentially contain a dozen distinct claims. Consequently, not every assertion they make will be backed with evidence. You can recognize unsupported claims in a passage because they lack any logical connections to other parts of the passage, whether through Logic keywords or implied through context. If these claims are uncontroversial (*the sky is blue, 2 + 2 = 4, the heart pumps blood*), then there really isn't a problem with leaving the claim unsupported, and you're not likely to be asked about it in a question. Watch out, though, when you find an unsupported claim that is not widely accepted—anything that either you don't believe personally or that you know that some people might not believe. Those are the types of unsupported claims most likely to be asked about in a question.

Empirical Evidence

Whenever an author appeals to experience, particularly in the context of scientific studies, he is using empirical evidence. While anecdotes, historical accounts, and case studies draw upon experience, these are limited in value because they represent only single cases. In other forms of empirical evidence, such as surveys, statistical analyses, and controlled experiments, variables can be isolated, and evidence can be gathered by looking at a wide swath of experience.

Experiments are most likely to be used as support in Scientific passages, although occasionally they will appear in a Historical passage or even a Philosophical passage. When you see them, be sure to identify the specific claim that they are being used to support because this is a common question. The example Psychology passage earlier used primarily this kind of support.

Logical Appeals

In Chapter 2 of *MCAT CARS Review*, logical appeals were introduced as appeals to *logos*, the Greek word from which *logic* is derived. If a claim is supported logically, the author makes deductions on the basis of how concepts are defined and related to one another. For example, the **contrapositive** is logically supported by the conditional claim that it comes from, as discussed in Chapter 6 of *MCAT CARS Review*.

Two other examples of logical appeals are analogies and the elimination of alternative possibilities. Analogical reasoning is employed whenever two things known to be alike in (at least) one respect are concluded to be alike in a different respect for which there may not be direct evidence. For instance, if an author argues that chimpanzees are more self-aware than other primates because they have larger brains, then she is likely making an analogy between chimpanzees and

Real World

In clinical medicine, the different types of research protocols can be placed in a hierarchy based on the strength of their evidence. Double-blind, randomized controlled studies give the strongest possible evidence, followed by systematic reviews (meta-analyses), cohort and case–control studies, case series, and expert opinion. Some of these research protocols are discussed in Chapter 11 of *MCAT Physics and Math Review*.

Bridge

Deductive (top-down) reasoning starts from a set of general rules and draws conclusions from the information given. This is different from inductive (bottom-up) reasoning, which seeks to create a theory via generalizations from specific instances. These problem-solving patterns are discussed in Chapter 4 of *MCAT Behavioral Sciences Review*.

humans—both have brains that are larger than other primates (for humans, considerably more so), but it's hard to gauge precisely how "self-aware" chimpanzees can be directly, so the analogy gives support where no other might be found.

The most classic example of eliminating the alternatives is the **reduction to absurdity** (*reductio ad absurdum*), in which the opposite of what the author is trying to prove is shown to have logical consequences that are ridiculous or even self-contradictory. Another example is when alternative hypotheses are ruled out by considering their weaknesses. Technically speaking, eliminating the alternatives only helps to support the author's position if all of the other possibilities are ruled out—this is easy in the case of a reduction to absurdity but extremely difficult when attempting to, say, explain the cause of a phenomenon.

Logical appeals are most likely to be found in Philosophical passages, like the earlier Ethics passage, where they predominate, although they do make limited appearances in the other types of passages as well.

Authority

Whenever an author draws on another person or text to support his claims, that author is appealing to authority. This is a way of making an appeal to *ethos*, as discussed in Chapter 2 of *MCAT CARS Review*. Occasionally, authors will try to demonstrate their own expertise and use that as a basis to bolster their other arguments, but far more often authors draw on outside sources, using paraphrases and direct quotations.

The level of support this offers varies depending on the credibility of the authority. **Primary sources**—for instance, a firsthand account of a historic event or the novel of a writer being discussed—provide the greatest level of support (short of having the experience itself). **Secondary sources**—that is, analyses and commentaries on primary sources (or other secondary sources)—are more dubious in value, and their value varies based on the expertise of the authority being cited. If the individual cited is discussing an area in which she has expertise (and is widely believed to have expertise), then a secondary source provides a significant level of support.

The appeal to authority is extremely common in Arts and Historical passages as in the examples earlier, and it may also be found in some Philosophical passages, especially those in religion.

Appeals to the Reader

Many authors will try to use their readers to help ground their arguments by beginning from starting points that their intended audiences are likely to share. For instance, as first addressed in Chapter 2 of *MCAT CARS Review*, whenever

> **Bridge**
>
> In Chapter 6 of *MCAT CARS Review*, we presented analogical reasoning as a set of two pieces of evidence and a conclusion:
> - **Evidence: K** and **U** share similar corresponding characteristics (K_1 and U_1, K_2 and U_2, K_3 and U_3, and so on);
> - **Evidence: K** possesses characteristic K_x;
> - **Conclusion: U** will possess similar characteristic U_x;
>
> where K is a well-established entity and U is an entity that is only partially understood.

MCAT Critical Analysis and Reasoning Skills

an author asks a rhetorical question, it's expected that the reader will have one specific response that the author had in mind when writing it (usually just *yes* or *no*). Also included in this broad category would be emotional appeals (appeals to *pathos*), in which charged language or colorful description is supposed to evoke particular responses from the audience.

Other times, authors try to rely upon the memories or imaginations of their readers. When authors support a claim by drawing on an everyday experience that the reader can recollect, or through a hypothetical situation or thought experiment, they're actually making the reader do most of the work for them. Whenever authors draw upon a relatively noncontroversial claim—something that everyone in the audience is likely to believe, something that will mesh with the reader's intuition—they are engaging in this kind of support. Note that such a claim still counts as support even if the audience members have starkly different reasons for believing it.

The thought experiment in the Ethics passage and the rhetorical questions in the Arts and History passages are examples of this type of appeal. The morally loaded language found in the latter is undeniably an appeal to *pathos*.

Faulty Support

If an author backs up a controversial claim with another claim that is similarly controversial, that author is simply using one assertion to back another. Unless that assertion is itself supported in some way (through one of the other types discussed here), then this kind of support is extremely weak at best. It's practically like leaving a claim unsupported.

Appeals in this category would include the wide variety of **specious reasoning**—arguments that superficially seem plausible but are actually flawed. Examples draw upon fallacies in reasoning like *ad hominem* attacks (attacking a person's character), making hasty generalizations from insufficient data, applying assumptions about a group to individual cases (stereotyping), and refuting straw-man positions (extreme versions of a view that no one seriously holds). It's relatively rare that such errors in reasoning appear in CARS passages, so if you notice one, expect to see a question on it.

7.3 Anticipating Questions

Let's return once again to the four mini-passages from the first section, but this time view them with an eye toward the kinds of questions that would probably be included with each on Test Day.

Bridge

Lawrence Kohlberg's theory of moral reasoning provides a great example of individuals sharing the same beliefs, but with very different reasons for holding these beliefs. The Heinz dilemma, presented in Chapter 6 of *MCAT Behavioral Sciences Review*, is a classic example: depending on whether a person's reasoning is preconventional, conventional, or postconventional, he may justify his answer (whether it is right or wrong for an individual to steal a high-profit margin medication he cannot afford for his terminally ill wife) with completely different logic.

An Arts Passage

One of the first examples of the ascendance of abstraction in 20th-century art is the Dada movement, which Lowenthal dubbed "the groundwork to abstract art and sound poetry, a starting point for performance art, a prelude to postmodernism, an influence on pop art…and the movement that laid the foundation for surrealism." Dadaism was ultimately premised on a philosophical rejection of the dominant culture, which is to say the dominating culture of colonialist Europe. Not content with the violent exploitation of other peoples, Europe's ruling factions once again turned inward, reigniting provincial disputes into the conflagration that came to be known by the Eurocentric epithet "World War I"—the European subcontinent apparently being the only part of the world that mattered.

The absurd destructiveness of the Great War was a natural prelude to the creative absurdity of Dada. Is it any wonder that the rejection of reason made manifest by senseless atrocities should lead to the embrace of irrationality and disorder among the West's subaltern artistic communities? Marcel Janco, one of the first Dadaists, cited this rationale: "We had lost confidence in our culture. Everything had to be demolished. We would begin again after the *tabula rasa*." Thus, we find the overturning of what was once considered art: a urinal becomes the *Fountain* after Marcel Duchamp signs it "R. Mutt" in 1917, the nonsense syllables of Hugo Ball and Kurt Schwitters transform into "sound poems," and dancers in cardboard cubist costumes accompanied by foghorns and typewriters metamorphosize into the ballet *Parade*. Unsurprisingly, many commentators, including founding members, have described Dada as an "anti-art" movement. Notwithstanding such a designation, Dadaism has left a lasting imprint on modern Western art.

Because this passage is relatively heavy on opinion, expect to see questions that require identifying what the author would agree or disagree with. So, for instance, it's clear this author has a negative opinion of European colonialism and warfare, but a positive opinion of Dada—which is not so surprising because this art is supposedly a rejection of those tendencies in European culture. A more difficult opinion to untangle might be the one at the very end: the author attributes the view to *many commentators* that Dadaism is *"anti-art,"* yet the author uses a Difference keyword (*Notwithstanding*) to signal disagreement with that conclusion, which is consistent with the author's use of the word *art* to refer to Dada, such as in the very first sentence.

This passage is also abundant in details, so expect Detail questions, described in Chapter 9 of *MCAT CARS Review*, which require you to comb through the passage searching for particular bits from the text.

MCAT Critical Analysis and Reasoning Skills

> **Bridge**
>
> Passages with a lack of support or argumentation are likely to include *Reasoning Beyond the Text* questions, which bring in a new element of information and ask you to apply the information in the passage to a new scenario (Apply questions) or ask how the new information would impact the passage (Strengthen–Weaken [Beyond the Text] questions). These question types are discussed in Chapter 11 of *MCAT CARS Review*.

What this passage lacks is much in the way of support, aside from quotations presumably coming from an art scholar, in the first paragraph, and an artist, *one of the first Dadaists*, in the second. Because you can still expect to see questions on reasoning accompanying any passage, it's quite likely that they'll bring in new elements to account for the relative dearth of evidence.

An Ethics Passage

The most prevalent argument against doctor-assisted suicide relies upon a distinction between *passive* and *active* euthanasia—in essence, the difference between killing someone and letting that person die. On this account, a physician is restricted by her Hippocratic oath to do no harm and thus cannot act in ways that would inflict the ultimate harm, death. In contrast, failing to resuscitate an individual who is dying is permitted because this would be only an instance of refraining from help and not a willful cause of harm. The common objection to this distinction, that it is vague and therefore difficult to apply, does not carry much weight. After all, applying ethical principles of *any sort* to the complexities of the world is an enterprise fraught with imprecision.

Rather, the fundamental problem with the distinction is that it is not an ethically relevant one, readily apparent in the following thought experiment. Imagine a terminally ill patient hooked up to an unusual sort of life support device, one that only functioned to prevent a separate "suicide machine" from administering a lethal injection so long as the doctor pressed a button on it once per day. Would there be any relevant difference between using the suicide machine directly and not using the prevention device? The intention of the doctor would be the same (fulfilling the patient's wish to die), and the effect would be the same (an injection causing the patient's death). The only variance here is the means by which the effect comes about, and this is not an ethical difference but merely a technical one.

This passage repeatedly uses an Opposition keyword, *distinction*, so expect to see questions requiring you to understand the basic conceptual difference between active and passive euthanasia. The author's characterization of this as *not an ethical difference but merely a technical one* at the end of the passage is particularly likely to be important because it includes an additional opposition: ethical *vs.* technical differences.

The numerous Logic keywords mean you can also anticipate questions that ask about the author's argumentative structure called Strengthen–Weaken (Within the Passage) questions. These are discussed in Chapter 10 of *MCAT CARS Review*. Notice the author's use of an appeal to the reader's imagination by means of a thought experiment—this is definitely ripe for questioning.

A Psychology Passage

There is no shortage of evidence for the existence of systemic biases in ordinary human reasoning. For instance, Kahneman and Tversky in their groundbreaking 1974 work proposed the existence of a heuristic—an error-prone shortcut in reasoning—known as "anchoring." In one of their most notable experiments, participants were exposed to the spin of a roulette wheel (specially rigged to land randomly on one of only two possible results) before being asked to guess what percentage of United Nations member states were African. The half of the sample who had the roulette wheel stop at 65 guessed, on average, that 45% of the UN was African, while those with a result of 10 guessed only 25%, demonstrating that prior presentation of a random number otherwise unconnected to a quantitative judgment can still influence that judgment.

The anchoring effect has been observed on repeated other occasions, such as in Dan Ariely's experiment that used digits in Social Security numbers as an anchor for bids at an auction, and in the 1996 study by Wilson *et al.* that showed even awareness of the existence of anchoring bias is insufficient to mitigate its effects. The advertising industry has long been aware of this bias, the rationale for its frequent practice of featuring an "original" price before showing a "sale" price that is invariably reduced. Of course, anchoring is hardly alone among the defective tendencies in human reasoning; other systemic biases have also been experimentally identified, including loss aversion, the availability heuristic, and optimism bias.

This passage focuses on a particular phenomenon, so you can expect to see questions asking about what the term *anchoring* means or for examples of it. Because the author never explicitly defines the word, and you have to figure it out from the experiments described, you're especially likely to see questions about it.

In addition, you should also anticipate questions asking about what the cited experiments actually show and what assertions they support. These questions would fall broadly into the *Reasoning Within the Text* category.

A History Passage

In 1941, an exuberant nationalist wrote: "We must accept wholeheartedly our duty and our opportunity as the most powerful and vital nation...to exert upon the world the full impact of our influence, for such purposes as we see fit and by such means as we see fit." If forced to guess the identity of the writer, many US citizens would likely suspect a German jingoist advocating for *Lebensraum*. In actuality, the sentiment was expressed by one of America's own: Henry Luce, the highly influential publisher of the magazines *Life*, *Time*, and *Fortune*. Luce sought to dub the 1900s the

> **Bridge**
>
> When a passage uses a term repeatedly without explicitly defining it, expect a question that asks you for this definition. These questions, appropriately called Definition-in-Context questions, are discussed in Chapter 9 of *MCAT CARS Review*.

"American Century," calling upon the nation to pursue global hegemony as it slipped from the grasp of warring Old World empires. As a forecast of world history, Luce's pronouncement seems prescient—but is it justifiable as a normative stance?

Not all of Luce's contemporaries bought into his exceptionalist creed. Only a year later, Henry Wallace, vice president under FDR, insisted that no country had the "right to exploit other nations" and that "military [and] economic imperialism" were invariably immoral. It is a foundational assumption in ethics that the wrongness of an act is independent of the particular identity of the actor—individuals who pay no heed to moral consistency are justly condemned as hypocrites. So why should it be any different for nation-states? In accord with this principle, Wallace proselytized for "the century of the common man," for the furtherance of a "great revolution of the people," and for bringing justice and prosperity to all persons irrespective of accidents of birth. Sadly, Wallace never had the chance to lead the United States in this cosmopolitan direction; prior to Roosevelt's demise at the beginning of his fourth term, the vice presidency was handed to Harry Truman, a man whose narrow provincialism ensconced him firmly in Luce's camp. And with Truman came the ghastly atomic eradication of two Japanese cities, the dangerous precedent set by military action without congressional approval in Korea, and a Cold War with the Soviet Union that brought the world to the brink of nuclear destruction.

When two opposing views are presented, as in this passage, you can be sure that you'll get questions asking you about one of the views or requiring you to contrast them. You could see straightforward questions asking you about who says what, about whom the author favors (as noted in our sample outline earlier, the author clearly favors Wallace over Luce and Truman), or about the assumptions implicit in each view.

> **Key Concept**
>
> Passages that include two or more differing opinions will usually be accompanied by questions that ask about the differences between or among those opinions, who holds which opinions, or the assumptions implicit in the various opinions.

The author's strong opinion is also ripe for questions, just as was the case in the Arts passage. Although many authors are fairly moderate, a substantial minority do take more overt positions, heavily employing Positive, Negative, and Extreme Author keywords. While you may not get an explicit question asking about the extreme view, you can at least expect to use your knowledge of the author's attitude to eliminate inappropriate answer choices.

Conclusion

And so our discussion of passages draws to a close. The skills that you've gained in learning how to dissect passages—from understanding rhetorical elements to using keywords and reading critically, from using the Kaplan Method for CARS Passages to analyzing argumentation and formal logic, from recognizing varieties of passages to anticipating questions—will serve you well not only on the MCAT but in medical school and as a physician. While reading a CARS passage, we use keywords to guide our reading and determine how information is put together; we draw inferences and set expectations for where the author is likely to go with a given argument; we anticipate the likely questions we'll be asked. In medicine, we must *listen* critically to our patients, determining how the complaints they describe fit together. *I've just been feeling "blah,"* your patient may say to you some day. *I feel tired, my skin's been dry, I've been putting on some weight, and I always feel cold.* From this information, you may infer that the patient likely has hypothyroidism. You then set expectations for some of the other signs and symptoms the patient may have, which guide your physical exam: Does the patient have a goiter (a swelling of the thyroid gland)? What about delayed deep tendon reflexes? Perhaps a slow pulse rate? And you anticipate the questions the patient will ask you—*What does this mean for me? Do I have to take medication for this? Does this put me at risk for anything else?* In the end, the *Critical Analysis and Reasoning Skills* section gets its name because it tests your ability not to comprehend dance theory, musicology, archaeology, and linguistics, but to understand how to analyze and reason through complex information delivered verbally.

Passages are only a part of the picture, though. While we need to read the passage to gain information, the real points come from the questions. In the next four chapters, we shift our focus to these questions, starting with the Kaplan Method for CARS Questions and then focusing on the major types of questions that fall into the three AAMC categories: *Foundations of Comprehension*, *Reasoning Within the Text*, and *Reasoning Beyond the Text*.

MCAT Critical Analysis and Reasoning Skills

CONCEPT AND STRATEGY SUMMARY

Varieties of Passages

- **Humanities passages** include topics from architecture, art, dance, ethics, literature, music, philosophy, popular culture, religion, studies of diverse cultures, and theater. Many passages in the humanities can be considered Arts passages or Philosophical passages.
    - **Arts passages** tend to include strong opinions, quotations, and descriptive language to illustrate examples.
    - **Philosophical passages** tend to be abstract and heavy on logic, focusing heavily on concepts and relations between them; they often appeal to the reader's memory or imagination.
- **Social Sciences passages** include topics from anthropology, archaeology, economics, education, geography, history, linguistics, political science, population health, psychology, sociology, and studies of diverse cultures. Many passages in the social sciences can be considered Scientific passages or Historical passages.
    - **Scientific passages** tend to include empirical studies and more subtle author opinions.
    - **Historical passages** tend to draw on historical events and quotations from sources alive at the time; they may include empirical studies or theoretical evidence, which can make them similar to the other passage varieties.

Support in Passages

- **Unsupported claims** are assertions that lack evidence. If the claim is uncontroversial, no evidence is generally needed. If the claim is controversial, however, the absence of evidence makes the claim questionable—and likely to be tested.
- **Empirical evidence** includes surveys, statistical analyses, and controlled experiments, although it can also include anecdotes, historical accounts, and case studies. Empirical evidence is most frequently used in Scientific passages.
- **Logical appeals** include formation of the contrapositive, analogical reasoning, and elimination of alternative possibilities.
- **Appeals to authority** include references to outside sources, paraphrases, and direct quotations. This technique is commonly used in Arts and Historical passages.
    - **Primary sources** give the greatest level of support and are first-hand accounts directly from the time period or situation being discussed.
    - **Secondary sources** provide less support and include commentaries on or explanations for primary sources (or other secondary sources).

- **Appeals to the reader** include rhetorical questions, emotional appeals (using charged language or colorful description to evoke an emotional response from the audience), and appeals to memory or imagination.
- **Faulty support** comes in many forms, but it includes attacking a person's character, making generalizations, stereotyping, and refuting straw-man positions. Faulty support is rare, but—when present—is often tested.

Anticipating Questions

- In general, it is important to anticipate the questions that will be associated with a given passage because doing so increases your speed and accuracy when moving through the question set.
- Passages that are abundant in details often contain Detail questions.
- Passages that use a new term often contain Definition-in-Context questions.
- Passages that are heavy on opinion often contain Inference questions.
- Passages that contain multiple opinions often contain questions that require distinguishing or classifying different opinions.
- Passages that are heavy on Logic keywords or empirical evidence often contain Strengthen–Weaken (Within the Passage) questions.
- Passages that lack support will often ask for this support through *Reasoning Beyond the Text* questions.

MCAT Critical Analysis and Reasoning Skills

WORKED EXAMPLE

Use the Worked Example below, in tandem with the subsequent practice passages, to internalize and apply the strategies described in this chapter. The Worked Example matches the specifications and style of a typical MCAT *Critical Analysis and Reasoning Skills* (CARS) passage.

Passage	Analysis
In *Prisoners Of Men's Dreams*, published in 1992, Suzanne Gordon argues that American feminism has lost sight of its original goal of transforming the world into a kinder, gentler place. Gordon deplores the sort of feminism that has triumphed instead: a cold, ruthless, "equal-opportunity" feminism, which aims for women's entrance into the masculine public world and their achievement by male standards of excellence.	Without giving any bias, the author introduces Gordon's argument that feminism in America has become *cold* and *ruthless*. A simple Label so far: **P1.** Gordon: American feminist change is bad; kind → ruthless
The heart of the book consists of excerpts from a hundred interviews with career women, who do a lot of complaining about fatigue and disillusion. Gordon's subjects comprise an unsurprising lot, given her presupposition of modern feminism's focus on successful women as products of overcoming male-centric and male-infused social and business structures. At the end, Gordon calls for a National Care Agenda that would make "caregiving" rather than competition the ultimate American value.	This is background support for paragraph 1 and simply outlines where Gordon got her data, and that she has an overall *caregiving* message. There is still no point of view from the author. A quick Label: **P2.** Gordon's call for change, ex: National Care Agenda
Suzanne Gordon is obviously an intelligent, sympathetic, and well-meaning person, but *Prisoners Of Men's Dreams* is a good example of the kind of sentimental, unlearned effusion that has become a staple of contemporary feminism and that most men rightly ignore. And who could blame them? Rallying for the propulsion of women in the public and private spheres through carefully played attempts at the pity point are bound to be met with stolid expression and silenced ears.	The author reveals her point of view: Gordon is smart and *well-meaning*, but she is one of the several *sentimental, unlearned* theorists. In short, to the author, Gordon is off base and thus is *rightly ignore[d]*. Do we know a lot about the author's specific objections? Not yet, but we expect more. A Label for this paragraph might be: **P3.** Auth: Gordon is well-meaning but off base

Passage	Analysis
Like so many American feminists, Gordon is completely out of her depth as a social analyst. Awkward, unintegrated quotes from Adam Smith and Woodrow Wilson are waved around to disguise her lack of familiarity with economics, history, and political science. Gordon's quote appropriations smack of the same short-sighted social phenomena that lead to "Keep Calm and Carry On" paraphernalia being plastered on the walls with complete disregard of manifest intent of the message.	As expected, there are more details about why the author dislikes Gordon's book, and the author lumps her in with the rest of the American feminists. The last paragraph stated that Gordon is *sentimental* and *unlearned*; this paragraph only focuses on that second bit—how she fails at educated *social analys[is]*. The author's charges are somewhat wild so far: *waved around*, *lack of familiarity*, *quote appropriations smack*, and *complete disregard*. Will the indictment get more specific? An appropriate Label for this paragraph is: **P4.** Auth: Gordon bad at social analysis
We are presented with the usual three-handkerchief, tear-jerker scenario about Big, Bad, Ugly America—that corrupt, empty, greedy society that all those wonderful, warm, benevolent people around the world look at with disgust. This point of view is the essence of chic these days among know-nothing feminists and the preening pseudoleftists who crowd our university faculties.	The author continues to use hyperbolic phrasing like *three-handkerchief tear-jerker* and *preening pseudoleftists* to set a strong opinionated tone. We might already expect to see a couple questions about the author's point of view. A Label for this paragraph is: **P5.** Auth: Anti-American sentimentality is just popular now
Well, let me tell you: as a child of Italian immigrants, I happen to think that America is the most open, dynamic, creative nation on God's green earth. As a scholar, I also know that it is capitalist America that produced the modern, independent woman. Never in history have women had more freedom of choice in regard to dress, behavior, career, and sexual orientation.	While there were already plenty of clues to the author's dislike of Gordon's argument, we finally get the author's personal thesis. Women are better off now than ever before, and all of the things Gordon railed against are actually what make America the best for feminists. However, we still don't have any concrete evidence for any of the author's many claims in that last sentence. Unless we get that, expect Strengthen–Weaken or Inference due to an incomplete argument. Label this as: **P6.** Auth: America actually best for women now

MCAT Critical Analysis and Reasoning Skills

Passage	Analysis
And yet, Gordon's insistence on defining women as nurturant and compassionate drove me up the wall. My entire rebellion as a child in the Fifties was against this unctuous, preachy stuff coming from teachers, nuns, and Girl Scout leaders. This drivel was not the path to supporting and empowering this woman.	This paragraph goes off track from the main argument, and may come up in the questions because it seems to be more a personal commentary. At the very least, it's continuing the author's tirade against Gordon, but similar to the last paragraph, the argument is based solely on the author's own experiences and interpretation. A Label for this paragraph is: **P7.** Author dislikes preaching nurturing, re: Gordon
This "transformative feminism" is just as repressive and reactionary as the "patriarchy" it claims to attack. Minerva save us from the cloying syrup of coercive compassion! What feminism does not need, it seems to me, is an endless recycling of Doris Day Fifties clichés about noble womanhood.	Similar to paragraph 5, we get the hyperbolic overt language like *Minerva save us* and *Doris Day Fifties clichés* to hammer home the strength of the author's opinion. The thesis is that feminism is regressing and misguided, and the goals of Gordon and likeminded feminists will only hurt the cause overall. A final Label: **P8.** Auth: Feminism regressing, compassion-mindedness is bad for cause

Here's a sample Outline and Goal for this passage:

P1. Gordon: American feminist change is bad; kind → ruthless

P2. Gordon's call for change, ex: National Care Agenda

P3. Auth: Gordon is well-meaning but off base

P4. Auth: Gordon bad at social analysis

P5. Auth: Anti-American sentimentality is just popular now

P6. Auth: America actually best for women now

P7. Author dislikes preaching nurturing, re: Gordon

P8. Auth: Feminism regressing, compassion-mindedness is bad for cause

Goal: To rebut Gordon's argument for compassionate feminism

7: Understanding Passages

Question	Analysis
1. The author of the passage would most likely claim that someone who did NOT agree with her view of feminism was:	This is an Apply question because we are asked for the author's Response to a new scenario. Noticing the word *NOT* in question stem, we could use elimination or rephrase it in order to get closer to a prediction. *Someone who [does] NOT agree with [the author's] view of feminism* is just a more convoluted way of asking for what the author thought about Gordon and similar feminists. In paragraph 3, the author derides these feminists as *sentimental* and *unlearned*. Both of these are excellent predictions.
A. independent.	We can eliminate **(A)** because the author identifies with and is supportive of the *modern, independent woman* in paragraph 6.
B. sentimental.	**(B)** is a perfect match and is the correct answer.
C. rebellious.	**(C)** is an Opposite as the author confidently declared her own *rebellion as a child* in paragraph 7.
D. matriarchal.	**(D)** is a Faulty Use of Detail because while Gordon claims to attack patriarchy, according to paragraph 8, she does not necessarily advocate matriarchy; and further, the author is not against matriarchy—she is simply against *the cloying syrup of coercive compassion* that feminists might think is matriarchy.

MCAT Critical Analysis and Reasoning Skills

Question	Analysis
2. In the passage, "transformative feminism" (last paragraph, first sentence) is used to mean:	Because this question gives a direct quote that needs interpreting, it must be a Definition-in-Context question. We expect to read the sentence before and after to get the gist, and also to confer with our Outline. Looking at the surrounding text, *transformative feminism* seems to be about *teachers, nuns, and Girl Scout leaders* applying Gordon's theory of womanhood—the one that our author has continually railed against as too *nurturant and compassionate*. In this sense, these typical "womanly" virtues are what the author believes America has successfully gotten away from, and it would be regressive to be forced to act that way. The prediction should be something like: *encouraging women to act in a nurturing and compassionate way.*
A. a political agenda with caregiving as its guiding principle.	(A) is a Distortion that describes Gordon's National Care Agenda. While this is an extension of Gordon's basic argument, *transformative feminism* describes the changes in women's roles, not political policy.
B. the process of reinventing power structures to provide equal opportunities for women.	(B) is an Opposite of Gordon's description of feminism; in paragraph 1, the author mentions that Gordon is against *equal-opportunity feminism*.
C. encouraging people to adopt "womanly" virtues as cultural standards.	(C) matches closely with the prediction and is the correct answer.
D. a goal of redefining feminine roles to include nurturing and compassion.	(D) is a tempting Distortion of the author's words. Gordon already sees the feminine role as nurturant, so the word *redefining* is flawed in this answer choice. In other words, Gordon claims that feminine roles already include nurturing and compassion—but that women should be more strongly encouraged to take on these roles.

7: Understanding Passages

Question	Analysis
3. Which of these assumptions is NOT implicit in the author's view?	We notice the *NOT* in the question stem, which makes this a Scattered Inference question (specifically of the Assumption subtype). These questions are often solved faster by elimination rather than predicting because the one correct answer could be nearly anything while the three wrong answers must necessarily have been part of the author's view. It's worth noting that the assumption we're looking for must offer support for one of the author's arguments, not simply something that could be true.
A. Stereotypes concerning female attributes are incorrect.	**(A)** can be eliminated because the author's central argument is anchored in her belief that Gordon's claim of womanly virtue is false.
B. "Equal-opportunity" feminism is fundamentally ideal.	At first, **(B)** seems like it could be inferred because the author argues against Gordon's definition of women as caring and compassionate throughout the passage. However, this does not mean that the author necessarily views *"equal-opportunity" feminism* as *ideal*. Indeed, it would not be surprising if the author noted some potential flaws with *"equal-opportunity" feminism* as well. Because **(B)** might be true—but does not have to be—it is *NOT* an *assumption…in the author's view* and is therefore the correct answer.
C. Gordon's rationale gives credence to Fifties-style clichés of feminine roles.	**(C)** is well-supported in the final paragraph when the author rallies against Gordon, stating that *feminism does not need…an endless recycling of Doris Day Fifties clichés about noble womanhood*.
D. American values and mores are considerably more liberal than those of other nations.	**(D)** is well-supported in paragraph 6, when the author states *I happen to think that America is the most open, dynamic, creative nation on God's green earth*.

MCAT Critical Analysis and Reasoning Skills

Question	Analysis
4. The author's most strenuous objection to Gordon's thesis is that:	Because this question asks about the relative strength of the author's objections, this is a Function question. The word *most* stands out because while there are many *objections* made, very few were actually backed up with any evidence besides the author's personal conviction. The *most strenuous objection* had better be defensible. The author refers to Gordon's work as *unlearned*, questions her writing and social analysis ability, and associates her opinions with *know-nothing[s]*. A prediction like, *Gordon's thesis is uneducated rambling* fits here.
A. it lacks scholarly rigor.	**(A)** closely matches with the prediction and is the correct answer.
B. it offers political solutions for moral questions.	**(B)** is Out of Scope because the author never mentioned morality or *moral questions* in the passage.
C. it depicts modern feminism as cold and ruthless.	**(C)** is an accurate description of Gordon's view of feminism but is not a significant objection of the author's.
D. the modern independent woman is free to choose her dress, career, and behavior.	The author would certainly agree with **(D)**, but it is not actually an objection against Gordon; in theory, Gordon wants those same things, just cast in a much different light.

7: Understanding Passages

Question	Analysis
5. The author's opinion of Gordon's view of American values can best be described as:	There are a few opinions being thrown around. Let's attack it piece-by-piece: What do we know about *Gordon's view of American values*? Gordon views them as misplaced, cold, or wrong, and wants to bring back nurturing and compassion as core values. Then, how does the author feel about that view? The author dismisses Gordon's efforts as *rightly ignored* and then challenges Gordon's views consistently throughout the passage. This is actually a masked Main Idea question, asking about the author's tone throughout the passage (which mostly focuses on the author's attacks of Gordon). Predict something negatively biased and opinionated.
A. disinterested.	**(A)** can immediately be eliminated. *Disinterested* is not a synonym for uninterested; rather, disinterested means unbiased—which certainly does not describe the author!
B. tolerant.	**(B)** is an Opposite; the author derides Gordon's views throughout the passage.
C. uninformed.	**(C)** is incorrect because the question stem asks for our description of the author's opinion—not the author's description of Gordon's opinion. The author herself could not be described as uninformed; if anything, the author hints at having much more experience and data than Gordon.
D. dismissive.	**(D)** matches perfectly to the negative and opinionated tone of the author.

Practice Questions

Passage 1 (Questions 1–5)

In recent years, extensive media attention has been given to enormous damages awarded in the US civil litigation tort system. In 1996, 79-year-old Stella Liebeck was awarded 2.7 million dollars in punitive damages from McDonald's after sustaining third degree burns from spilled coffee. The system awarded Michael Gore nearly four million dollars in 1994 after BMW sold him a car that had been repainted and sold as new.

Awards such as these spurred businesses, insurance companies, and lobbyists to claim an "explosion" of legal liability. In response, many legislators call for tort system reform that includes limiting the amount of damages, controlling legal fees, and redefining the concept of "fault" administered by the judges. Jury verdicts that appear, on superficial inquiry, to be blatantly excessive seem to challenge our system of compensation. Some claim that juries find negligence in order to provide compensation for victims who have large medical bills and lost wages, at the expense of "deep pocket" defendants.

In his seminal article in the *Maryland Law Journal*, "Real World Torts: An Antidote to Anecdote," Marc Galanter examines the issue. As the title suggests, in order to investigate the tort system, Galanter used empirical data to examine whether, on the whole, these "anecdotes" truly represent how the system compensates injured parties.

Galanter found that all tort claims form a dispute pyramid charting the progress from an injury to a jury verdict. Injuries form the broad base of the pyramid. On the next level, approximately 8% of injuries become grievances (events for which an injury was noticed). Of these grievances, 85% become claims (where the injured brings the problem to the alleged wrongdoer), and 23.5% of claims become disputes (having failed to reach an informal agreement). Next, 58% of plaintiffs with claims contact a lawyer, and 32.8% of these result in a court filing. Of all court filings, only 7% result in a verdict, and only 34.7% of these are decided in favor of the plaintiff. This means that an injured person gets a jury verdict in his favor only 0.007% of the time.

For example, medical malpractice results in approximately 100,000 deaths a year. At the tip of Galanter's pyramid only 21 of the 100,000 deaths will result in a verdict. Finally, only 7 people will receive damage awards from a jury.

Galanter concludes that the system is hardly unbalanced in favor of plaintiffs. The proposed tort reform would actually increase insurance company profitability and reduce payments to the most seriously injured tort victims. Punitive damage awards are extremely rare, only applied in the most egregious cases, and always subject to judicial review. The awards discourage businesses from releasing harmful products into the stream of commerce.

Moreover, according to Galanter, court filings in the law division of the circuit court of Cook County have actually declined during the period from 1980 to 1994. His observations are consistent with a 1999 study by the National Center for State Courts, which found that tort filings have decreased by 9% since 1986. By looking at existing empirical data instead of isolated, inflammatory cases, legislators will be able to do a better job of deciding if the system is in need of reform and, if so, what type of reform is appropriate.

1. The author primarily mentions Liebeck's award (paragraph 1) in order to:
   A. give an example to support his overall claim.
   B. give an example of a verdict that is blatantly excessive.
   C. give an example of a verdict that has caused legislators to call for tort reform.
   D. introduce evidence for a conclusion made later in the passage.

2. Which of the following situations would be most analogous to the situation faced by a potential tort plaintiff, based on the information in paragraph 4?
   A. A young basketball prospect trying to make it to the NBA
   B. A group of children picking sides for a baseball game
   C. A young, qualified woman looking for a job
   D. An injured woman trying to reach an emergency room

3. Based on the information in the passage, the author believes that Galanter's pyramid:
   I. is applicable to medical claims.
   II. should compel legislators to change their views.
   III. is biased against "deep pocket" companies.
   A. I only
   B. II only
   C. I and II only
   D. I, II, and III

4. Which of the following would most WEAKEN the conclusion implied in paragraph 4?
   A. Galanter's article was published in 1995.
   B. A study that shows filings for divorce following a much different pattern.
   C. The fact that Galanter's study dealt with only product liability cases.
   D. Most doctors carry medical malpractice insurance.

5. Based on the information in the passage, the author would argue for all of the following EXCEPT:
   A. juries should not be so compassionate toward victims at the expense of wealthy businesses.
   B. legislators should examine all data.
   C. the media spotlight does not necessarily clarify problems.
   D. courts might award damages as a way to ensure that businesses practice in the public's best interest.

Passage 2 (Question 6-12)

...[post-World-War-II Director of Policy Planning George F.] Kennan's strategy had been to try to bring about changes, over time, in the Soviet concept of international relations: to convince Russian leaders that their interests could be better served by learning to live with a diverse world than trying to remake it in their own image. Kennan had rejected both war and appeasement to accomplish this; it could only be done, he thought, through a long-term process of "behavior modification"—responding positively to whatever conciliatory initiatives emanated from the Kremlin, while firmly countering those that were not...

Kennan took the position that it was as important to reward the Kremlin for conciliatory gestures as it was to resist aggressive ones. This meant being prepared to engage in such negotiations that would produce mutually acceptable results. The [Truman] administration conveyed the appearance of being willing to discuss outstanding issues with Moscow, but Kennan regarded several of its major actions between 1948 and 1950...[among them] the formation of the North Atlantic Treaty Organization (NATO)...as certain to reinforce Soviet feelings of suspicion and insecurity, and hence, to narrow opportunities for negotiations...

The initiatives for the North Atlantic Treaty came from the Western Europeans themselves, and reflected the uneasiness they felt over the disparity in military power in Europe: the Russians had thirty divisions in Eastern and Central Europe alone; comparable combined US, British, and French forces came to fewer than ten divisions. Thus, the Western Union countries (Great Britain, France, and Benelux), together with the United States and Canada, agreed on the outlines of a treaty providing that an attack on any one of them would be regarded as an attack upon all.

Kennan had not been involved in the initial discussions, but he made clear his reservations about the course the administration chose to follow. These boiled down to three points: (1) that the Europeans had mistaken what was essentially a political threat for a military one, and that they consequently risked "a general preoccupation with military affairs, to the detriment of economic recovery"; (2) that outside the immediate North Atlantic, "which embraces a real community of defense interest firmly rooted in geography and tradition," any alliance extended to only some countries would render the rest more vulnerable...; (3) that an alliance made up of [Western European] nations would amount to "a final militarization of the present dividing-line through Europe," and that "no alteration, or obliteration, of that line could take place without having an accentuated military significance." Such a development might be unavoidable, "but our present policy is still directed...toward the eventual peaceful withdrawal of both the United States and the U.S.S.R. from the heart of Europe..."

These were not isolated concerns. There was worry in Washington that emphasis on rearmament would delay recovery; indeed, one condition attached to the administration's military assistance program for Western Europe was that economic revival would continue to have first priority. The question of how to include some countries without appearing to write off others also caused a great deal of agonizing: in the end, the administration stretched the concept of "North Atlantic" to encompass Italy, but refused to extend it to Greece, Turkey, Iran, or to form a comparable pact with non-communist countries of the Western Pacific. There was less concern about Kennan's third point simply because most observers already regarded division, by mid-1948, as an accomplished fact...Despite its reservations, the administration went on to conclude a North Atlantic Treaty and initiate a program of military assistance to its members. Kennan came to see, regretfully, that [because of military insecurity of Europeans] there were few alternatives...

6. The passage suggests that Kennan's "behavior modification" approach to changing the Soviet concept of international relations was:

   A. unlikely to be successful if the Kremlin always made conciliatory gestures.
   B. moderate in comparison with the approaches he decided to reject.
   C. a logical outgrowth of his extensive background in behavioral psychology.
   D. an extension of American strategy during World War II.

7. The passage suggests that the impetus for the formation of NATO was:

   A. information that a Russian attack on Western Europe was impending.
   B. the understanding that no nation could withstand a Russian attack without assistance.
   C. the desire to aid the Western European economic recovery as well as to guarantee military assistance.
   D. the fear that the Soviets would try to capitalize on their military advantage.

8. Kennan assumed which of the following in making his first counterpoint against NATO?

   A. The formation of the military alliance would spur economic growth.
   B. The presence of the thirty Soviet divisions did not mean they were going to attack.
   C. The economic recovery in Europe had been progressing slowly.
   D. It's always a mistake to make military affairs a higher priority than economic affairs.

9. Kennan's reaction to the administration's refusal to extend NATO membership to Greece, Turkey, or Iran was most likely one of:

   A. understanding because these countries did not have the same geographic defense interests as the Europeans.
   B. approval of the fact that the concept of "North Atlantic" was not overextended.
   C. disappointment that those countries could not now be employed in anti-Soviet strategy.
   D. trepidation that these countries were now more open to potential enemy aggression.

10. Which of the following explains why the Truman administration was not worried about Kennan's objection that NATO would amount to "a final militarization of the present dividing-line through Europe"?

   A. They believed that it would be possible to alter the line through negotiations of peaceful withdrawal.
   B. They wanted to maintain a strong American military presence in Europe.
   C. They felt it was too late to prevent the solidification of the dividing line.
   D. Neither of Kennan's other two objections to NATO had given them cause for concern.

11. The passage suggests that, with regard to the reservations expressed by Kennan about NATO, the administration was:

   A. often in agreement but ultimately undeterred.
   B. unresponsive to his proposals for improvement.
   C. able to counter each of his criticisms.
   D. forced to carefully re-examine its objectives.

12. Which of the following conclusions would be most in accord with the theme of the passage?

    A. Military alliances invariably have drawbacks that render them ineffective.
    B. Behavioral modification is the only way to change a government's concept of international relations.
    C. Coherent international strategy can flounder because of the military situation.
    D. Negotiations should be conducted between two powers once military equilibrium has been established.

Explanations to Practice Questions

Passage 1 (Questions 1–6)
Sample Passage Outline

P1. Introduction to tort system; examples of extreme damages awarded

P2. Insurance/lobbyists: tort awards too excessive

P3. Galanter examines empirical evidence

P4. Galanter pyramid findings: tiny amount of successful tort claims

P5. Malpractice ex: 7 of 100,000 win damages

P6. Galanter: system not in favor of plaintiffs; awards = good: keeps businesses in line

P7. Overall, tort filings declining without legislation, Auth: legislators should use empirical data

Goal: To examine the tort system and the proposed legislation against it, and to outline Galanter's clear rebuttal of those arguments

1. C

This Function question should direct us to our Outline. Paragraph 1 mentions extreme examples the tort system awards. Even without rereading, predict that Liebeck must be one of those very high monetary awards that caused lobbyists to be so aggressive in fighting against the tort system in general and call for reform. This matches to **(C)**. **(A)** is not possible simply because the author does not overtly make any claims. Our author is very neutral and simply sets the facts in front of us to make our own decisions. **(B)** distorts the author's mention of *jury verdicts that appear, on superficial inquiry, to be blatantly excessive*—the use of the phrase on superficial inquiry implies that the author may not agree that these damages actually are excessive. Finally, **(D)** is vague enough that it could sound plausible, but there is no later argument or conclusion that requires Liebeck's award specifically. Generally, there isn't any conclusion about coffee, McDonalds, burns, or the elderly, that would depend on this example either.

2. A

For this Apply question, our Outline is an excellent place to find a prediction. The best description of *the situation faced by a potential tort plaintiff* is found in paragraph 4: Galanter's pyramid findings demonstrated that only a tiny number of tort claims actually result in decisions in favor of the plaintiff. The main point is that there must be a tiny number of success stories from a much larger pool of individuals. This matches to **(A)**. In each of the other cases, close to 100% success would be expected (or at least a much higher percentage than the number of young basketball prospects who make it to the NBA).

3. C

The combination of *Based on the information in the passage* and Roman numerals in the question stem tells us that this is a Scattered Detail question. Turning to our Outline for paragraphs 4 and 5, the author brings in the medical example to illustrate the extremely low rate of success that actually is seen in tort cases. This is not a blatant endorsement of

MCAT Critical Analysis and Reasoning Skills

Galanter, but should coincide neutrally with Galanter's pyramid findings. Statement I is exactly the example that is being used, so eliminate **(B)**. Statement II is true as it is the recommended course of action given at the end of the passage, which is primarily supported by Galanter's pyramid argument. With **(A)** eliminated, we can investigate Statement III. This claim is actually made by lobbyists and insurance companies in paragraph 2—and is part of the larger claim that Galanter ultimately refutes. Thus, Statement III is untrue, making **(C)** the correct answer.

4. C

This is a Strengthen–Weaken (Beyond the Text) question, so let's start by determining the conclusion implied by paragraph 4. The main point of Galanter's argument stems from the assumption that analyzing all of the empirical data will give the fullest picture and not allow anecdotal bias. If it were possible that the data Galanter used was false or incomplete for some reason, this would seriously weaken his argument overall. **(C)** details exactly that prediction; if Galanter's study was only specific to product liability, then it can't be generalized to other similar cases or other tort suits. **(A)** reflects on the dates given in the passage—Galanter's study seems to investigate data until 1994. As long as Galanter's study was published after this point, there is no negative effect on his argument, eliminating this answer. The pattern of divorce cases, as described in **(B)**, has no effect on the argument because there is no reason to believe that divorce would (or wouldn't) follow the same patterns as tort cases. Finally, whether or not physicians have malpractice insurance does not appear to be related to the number of cases brought to court or decided in favor of the plaintiff, so **(D)** would also have no effect on Galanter's argument.

5. A

We are asked for *arguments* the author would make *based on the ... passage*; with the word EXCEPT included, this must be a Scattered Inference question. While the author does not overtly state an opinion, it can be inferred that the author sides overall with Galanter's thesis. With that as a general

prediction, we Plan to eliminate any answer choice that fits with Galanter (and, by extension, our author)—and the one answer that does not fit is correct. This means that **(A)** is immediately correct. The author would not openly support being more lax to wealthy businesses at the potential detriment to victims. This position is exactly Opposite Galanter's, making it the correct answer. The other claims are all made in the passage: **(B)** is supported by the claim *By looking at existing empirical data, ... legislators will be able to do a better job of deciding if the system is in need of reform*. **(C)** is supported by the *extensive media attention* described in paragraph 1. **(D)** is supported in paragraph 6, where the author states that *awards discourage businesses from releasing harmful products into the stream of commerce*.

Passage 2 (Question 6–12)
Sample Passage Outline

P1. Kennan: use behavior modification to improve Soviet international relations

P2. Behavior modification requires positive and negative reinforcement, ex: Truman administration

P3. NATO = Western nations uneasy, think Soviet military too big: attack on 1 = attack on all

P4. Kennan's reservations about NATO: (1) political *vs.* military threat, (2) other countries vulnerable, (3) solidifies military line in Europe

P5. Administration agrees with Kennan's 1 and 2, assumes 3 is already fact; NATO still goes through

Goal: To explain Kennan's reservations about NATO, and explain how and why it negatively impacted Soviet relations

6. B

Because this question is asking about something a conclusion the author implies, this is an Inference question of the Implication subtype. Our Outline for paragraph 1 mentions that *"behavior modification"* was Kennan's advocated method to improve international relations with

158

Soviets. It relied on consistent positive and negative responses, and not on military or extreme actions. This method is in contrast to the military strategy presented later in the passage, NATO—which Kennan opposed. Predict that Kennan's approach is more subtle and less aggressive than the other option. This matches closely to **(B)**. **(A)** isn't well-supported by the passage, and it comes across as an Opposite. *Conciliatory gestures* would mean that the Soviets are taking actions that the Western nations approve of, so this would theoretically mean that great progress (success) was underway. **(C)** ascribes an extensive background to Kennan that was never mentioned (or even hinted at) in the passage. Finally, **(D)** has two issues: first, American, World-War-II strategy was not explicitly described in detail (to determine if this could be an offshoot from that), and second, it appears that Kennan's *"behavior modification"* strategy wasn't actually used by America—against Kennan's reservations.

7. D

This is another Inference question, so we start by identifying the relevant text in the passage using our Outline. The most sensible place to go is the paragraph where NATO is first mentioned, paragraph 3. The suggested Label for this paragraph is: *NATO = Western nations uneasy, think Soviet military too big*. Specifically, Russia had more divisions than combined Western forces, and the West feared a possible (but entirely theoretical) attack. This matches closely with **(D)**. **(A)** intones a similar idea, but is too Extreme and Distorts the information. The attack was possible, but there was no evidence that *information* was available that showed the attack was *impending* (about to happen). **(B)** is similarly Extreme; even though certain Western European countries feared an invasion (and the passage hints at the dire fate of even smaller countries), the idea that *no nation could withstand a Russian attack* is simply too broad. Finally, **(C)** contradicts both Kennan and the administration, who say in the passage that *emphasis on rearmament would delay recovery*. Thus, while NATO did guarantee military assistance, it went directly against the principle of *aid[ing] Western European economic recovery*.

8. B

The word *assumed* tells us that this is an Inference question of the Assumption subtype. Our task is to find a missing piece of evidence for Kennan's first conclusion in paragraph 4. Kennan accused *the Europeans [of] mistak[ing] what was essentially a political threat for a military one*. One way to attack an Assumption question is the Denial Test, described further in Chapter 10 of *MCAT CARS Review*. In the Denial Test, one takes the opposite of each answer choice—whichever answer choice, when turned into its opposite, destroys Kennan's argument will be the correct answer. The opposite of **(A)** would be: *The formation of the military alliance would not spur economic growth*. This is a statement Kennan is likely to agree with, given the statement that *emphasis on rearmament would delay recovery* in paragraph 5, so it can be eliminated. The opposite of **(B)** would be: *The presence of the thirty Soviet divisions meant they were going to attack*. This destroys Kennan's argument because it means that the Europeans were not just responding to a political threat—there was a very real military threat as well. Because this negated statement ruins Kennan's argument, this must be the correct answer. Negating **(C)** would yield: *The economic recovery in Europe had been progressing quickly*. This assumption does not impact Kennan's argument and can be eliminated. Finally, **(D)** can be eliminated without even using the Denial Test because it is too Extreme. While Kennan might agree with this statement as it pertains to this particular situation and NATO, we do not have enough information to extrapolate to all other cases.

9. D

This is yet another Inference question, asking for Kennan's likely response. In summarizing Kennan's second point of contention, we Outlined: *(2) other countries vulnerable*. Kennan is worried that *any alliance extended to only some countries would render the rest more vulnerable*. **(D)** paraphrases that prediction. Both **(A)** and **(B)** express a neutral to positive reaction, which does not match Kennan's actual response. **(C)** might sound reasonable in a real-world scenario, but Kennan was generally against uniting formally against Russia. It's doubtful, then, that he would be *disappoint[ed] that those countries could not now be employed in anti-Soviet strategy*.

10. C

This is a Detail question asking for the Truman administration's thoughts on Kennan's final point of contention in paragraph 4. Paragraph 5 is where we learn what the administration's feelings actually are. The end of that paragraph serves as an excellent prediction: *most observers already regarded division, by mid-1948, as an accomplished fact*. This matches (C). (A) is a Faulty Use of Detail as it ascribes one of Kennan's opinions from the end of paragraph 4 to the Truman administration. (B) Distorts the description of the troops in Europe. The passage discusses the Western European troops that are present, but doesn't mention how many—or whether there even are—American ones there. Further, there is no evidence in the passage that the Truman administration desired *a strong American military presence* in Europe at all. (D) is an Opposite: the administration agreed with Kennan on the first two points (mostly), giving them *cause for concern*.

11. A

This Inference question is quite similar to the Detail question that preceded it. We know from paragraph 5 that the Truman administration agreed with Kennan on two of his three points, but that they still decided to go ahead with the treaty *despite [his and their] reservations*. This prediction matches nicely with answer (A): the administration was *often in agreement* with Kennan but *ultimately undeterred* by the reservations that he had. (B) indicates that Kennan had specific *proposals*, but they were never mentioned or implied in the passage. Kennan had general strategies like *"behavior modification"* but not specific proposals. (C) is patently untrue because the administration actually agreed with the majority (two out of three) of Kennan's points. Finally, (D) was also never directly discussed, and—if anything—the administration went ahead with actions to further *military assistance* and *economic revival* (the only two objectives that we're told about in the passage).

12. C

This is an excellent chance to turn the Goal into a prediction: *To explain Kennan's reservations about NATO, and explain how and why it negatively impacted Soviet relations*. Kennan's *"behavior modification"* strategy was hampered by the formation of NATO, and NATO was formed in response to the military imbalance in Europe—overall, the development of NATO was not beneficial to international relations with Russia. This matches closely with (C). (A) and (B) are too Extreme: NATO had *drawbacks*, sure, but was not *render[ed]... ineffective* by them. And behavioral modification is one—but not the *only way*—to change a government's concept of international relations. (D) is what the members of NATO clearly thought, but it is neither a main theme of the passage nor what Kennan—the primary voice in the passage—thought.

Question and Answer Strategy

8

8: Question and Answer Strategy

In This Chapter

8.1 The Kaplan Method for CARS Questions	**164**	**8.3 Signs of a Healthy Answer**	**170**
Assess	165	Appropriate Scope	171
Plan	165	Author Agreement	172
Execute	167	Weaker Is Usually Better	172
Answer	167	**Concept and Strategy Summary**	**173**
8.2 Wrong Answer Pathologies	**168**	**Worked Example**	**175**
Faulty Use of Detail (FUD)	169		
Out of Scope (OS)	169		
Opposite	170		
Distortion	170		

Introduction

> **LEARNING GOALS**
>
> After Chapter 8, you will be able to:
>
> - Attack questions by applying the Kaplan Method for CARS Questions: Assess, Plan, Execute, and Answer
> - Recognize and avoid common wrong answer pathologies
> - Locate correct answers by focusing on scope, author agreement, and tone

Thus far, this book has examined the multifarious aspects of *Critical Analysis and Reasoning Skills* (CARS) passages. You've learned about their rhetorical characteristics in Chapter 2 and their logical structure in Chapters 5 and 6. You've seen how to approach them strategically, using the keywords introduced in Chapter 3 and following the Kaplan Method for CARS Passages elaborated upon in Chapter 4. In Chapter 7, these threads were brought together as we discussed what understanding passages really means.

We now begin to consider question stems and answer choices. We start by outlining the Kaplan Method for MCAT Questions, as tailored to the CARS section. Subsequently, we look at the recurring traps that the testmakers set for the unwary student, which we call Wrong Answer Pathologies. In the final portion, we'll consider the counterpart to pathologies: patterns common in correct answers.

8.1 The Kaplan Method for CARS Questions

In Chapter 4 of *MCAT CARS Review*, we saw how the general Kaplan Method for tackling MCAT passages could be refined for the needs of the CARS section. In this section, we'll do the same with our question method, which takes the basic form shown in Figure 8.1.

ASSESS THE QUESTION

PLAN YOUR ATTACK

EXECUTE THE PLAN

ANSWER BY MATCHING, ELIMINATING, OR GUESSING

Figure 8.1. The Kaplan Method for Questions

This same four-step approach should be used on all questions on the MCAT—in both the *Critical Analysis and Reasoning Skills* section and the science sections. The CARS-specific version is shown in Figure 8.2.

ASSESS THE QUESTION
- Read the question, **NOT** the answers
- Identify the question type and difficulty
- Decide to attack *now* or *later*

PLAN YOUR ATTACK
- Establish the task set by the question type
- Find clues in the stem on where to research
- Navigate the passage using your Outline

EXECUTE THE PLAN
- Predict what you can about the answer
- Set expectations for wrong choices
- Be flexible if your first Plan flops

ANSWER BY MATCHING, ELIMINATING, OR GUESSING
- Find a match for your prediction, or
- Eliminate the three wrong options, or
- Make an educated guess

Figure 8.2. The Kaplan Method for CARS Questions

Note: The Kaplan Method for CARS Questions, as well as the Kaplan Method for CARS Passages, CARS Question Types, and Wrong Answer Pathologies are included as tear-out sheets in the back of this book.

8: Question and Answer Strategy

ASSESS

You might notice that the first step of the question method is similar to the Scan step of the Kaplan Method for CARS Passages. This is not a coincidence but rather a consequence of the timing constraints posed by the section. Because every question is worth the same number of points, there's no reason to get derailed by any one of them. Be honest with yourself: at least a few questions on each section are so difficult that you're likely to get them wrong no matter how many minutes you spend on them. Wouldn't it be better to recognize which questions those are right away so you can instead use that precious time where it will actually pay off?

Toward that end, your first task with any question will be to read the stem, and only the stem, for the sake of deciding either to work on it *now* or to triage (to use an apt medical metaphor) and save it for *later*. Assessing the difficulty will be easier if you can identify the **question type**. The three large **categories** that the AAMC uses were discussed in Chapter 1 of *MCAT CARS Review*, but through extensive research of all released MCAT material, we've discovered that almost all *Foundations of Comprehension* questions fall into one of four types. We've also found that the *Reasoning Within the Text* and *Reasoning Beyond the Text* categories can each be split into two predominant types with assorted others appearing rarely. Determining the question type will make devising and Executing a Plan much easier, which is why we devote the next three chapters to the types and their unique strategic concerns.

Why avoid looking at the answer choices? The primary reason is that most of them are wrong. If you glance at just one of them, for instance, it's three times more likely to be incorrect than correct and could seriously mislead you about the question. Inexperienced test takers immediately jump to the answers, and the AAMC punishes them for it by wording wrong options seductively. Selecting the first answer that looks good without first formulating a Plan is just a recipe for failure. Thus, until you get in the habit of ignoring the answers entirely until the Answer step, use your hand or a sticky note to cover them up whenever you start to work on a question.

PLAN

Once you've decided to attack a question, it's time to Plan your attack. First, be clear about what the question requires of you—what we call the **task**. Simpler question types like Main Idea and Definition-in-Context always involve one specific task (recognizing the big picture and explaining the meaning of part of the text as used in the passage, respectively), and even the more complex question types have only a small number of common objectives. An example is the category of Apply questions, almost all of which involve one of three tasks: gauging the author's response, predicting a likely outcome, or finding a good example (as described in Chapter 11 of *MCAT CARS Review*).

> **Real World**
>
> Think of question typing as being like presenting diagnoses. In medical school, you will learn the appropriate steps to undertake if a patient presents with chest pain, abdominal pain, a severe headache, confusion, and so on. The ten question types should be thought of as the "presentation" of the question, which determines the appropriate steps to take through the rest of the Kaplan Method for CARS Questions.

> **MCAT Expertise**
>
> To avoid getting seduced by wrong answer choices, cover all answer choices with your hand or your scratch paper; leave them out of your field of view until you're ready for the Answer step. This will save you significant time and effort because you will avoid putting active consideration toward (and possibly being misled by!) the three wrong answer choices. Reading the answer choices can lead to the misinformation effect, discussed in Chapter 3 of *MCAT Behavioral Sciences Review*.

In the journey to a correct answer, the task is your destination. To get to where you're headed, you need to know where you're starting from. Thus, the other major aspect of the Plan step is to determine where you'll find the information that you need to Answer the question. Note that there are only four viable sources of information on Test Day: the *question stem*, the *passage*, your *Outline*, and your *memory*.

It's rare, but occasionally you'll find that the stem gives you everything you need to answer the question. In other cases, the relevant info might be fresh in your mind, although human memory is notoriously faulty, and it doesn't hurt to check another source. This is one reason why it's so valuable to create an Outline as you read; it serves as a memory aid, capturing important aspects of the text that could slip your mind. In fact, sometimes the Outline is the only place you'll need to check, saving you the trouble of researching the passage. For instance, with Main Idea questions, the author's Goal in your Outline is usually all you'll need to examine. Not every question in the CARS section requires rereading the text: if a simple Plan yields immediate success, accept it and don't second-guess yourself.

Whenever you're facing less straightforward tasks, though, it's advised that you reread at least part of the passage. Stems almost always contain clues hinting at where to research, taking one of three forms:

- Paragraph or sentence references
- Direct quotations
- Paraphrases

Sometimes paragraph numbers will be listed in parentheses, or you'll see a phrase like *the final sentence of the passage*, which makes location a breeze. The other clues may be harder to work with. Keep in mind that not all quotations taken from the passage will be surrounded by quotation marks, particularly if only a single word or short phrase is mentioned; these brief terms are sometimes italicized but may appear without any embellishment at all. Also, keep in mind that shorter quotations could actually appear in multiple paragraphs. Although rare, mercifully, paraphrases of passage claims without additional clues do occur and can be especially tricky: always remember that stems can use synonyms instead of the author's original language. A good Outline will help you find the paragraph that contains the information when you are working from a less direct clue.

Once you've figured out where to go in the passage, you'll want to make sure you cast a large enough net to find what you need. While the necessary information is often found directly in the portion of the passage that the stem points to, not infrequently you will find that the answer is in the surrounding text, in sentences immediately before or after. As a general rule of thumb, Plan to start rereading one sentence before the reference and to stop one sentence after.

MCAT Expertise

If you ever find yourself Scanning the passage to find an answer choice, you're doing it wrong! The information you need for the answer is often in your Outline; even if not, using your Outline to determine where in the passage to look for the answer means that you'll usually Scan—at most—one paragraph of text.

What should your Plan be if you can't locate the clues or need to contend with vague questions that lack them entirely? Typically such questions are best to try later, at the very end of a passage set, after you've researched the other questions and already reread some of the text. You may find that by the time you return to it, the effort you put into other questions ended up revealing an unhelpful question stem's correct answer. When you do attempt these questions, process of elimination will usually end up being the best Plan.

EXECUTE

Every question has exactly one correct answer but typically will have more than one way to reach it. Indeed, the only difference between discovering the one correct answer and ruling out the three wrong ones is that the latter approach usually takes more time; the result is identical. Consequently, you want to strive for the quickest approach, but be flexible so that if Plan A fails, you have Plans B and C to fall back on.

In the Plan step for most questions, you'll identify where you need to reread (in your Outline or in the passage itself). Keep in mind the task of the particular question type as you now Execute the Plan and look over your Outline or the text again. Here, unlike with your first strategic reading of the passage (discussed in Chapter 4 of *MCAT CARS Review*), it's okay to reread the crucial sentences a few times if you need to because points are now on the line. In most cases, Plan A is to use the text to predict everything that you can about the correct answer, while Plan B is to set analogous expectations about what would rule out choices. The more specific these dual expectations, the easier it will be to isolate the answer.

For greater specificity in Planning and Executing an attack, see the strategy portion of the discussion for each of the question types covered in Chapters 9 through 11 of *MCAT CARS Review*. Where you look for information, as well as the way in which you approach rereading (including which keywords you pay attention to), will vary depending on the task.

ANSWER

All that remains now is to select the correct Answer. Scan the choices, and if you see an item that closely resembles what you expected, reread every word of that answer choice carefully to make sure it says precisely what you think it does. Then, select it and move on to the next question. At that point, *reading the other choices will not be worth your time*—be confident that you've Answered the question when you find a near-perfect match. When a good match is not available, *then* elimination is always an option.

That said, don't feel that you immediately need to resort to process of elimination if you read one answer choice and discover that the correct answer is going to take a completely different form than you first predicted. Part of being flexible is being

MCAT Critical Analysis and Reasoning Skills

able to revise your initial Plan, to set new expectations if the answer choices point you in that direction. The answer choices could technically be considered a fifth source of information, but keep in mind that they include a lot of misinformation and so should be treated with caution.

For example, a question stem may ask for a claim that undermines the author's thesis, a classic Strengthen–Weaken (Within the Passage) task, described in Chapter 10 of *MCAT CARS Review*. For such a question, you may go into the answer choices expecting that the answer is a refutation mentioned in the passage. As soon as you read choice (A), though, you discover it's something entirely new, never even hinted at by the author. As you'll learn in Chapter 11 of *MCAT CARS Review*, such questions fall into the related Strengthen–Weaken (Beyond the Passage) type instead. Before moving on to read choice (B), you should think about whether choice (A) answers the question in a way you didn't anticipate. If so, use it to help you alter your expectations. Avoid the temptation to abandon the critical thinking you performed during the Execute step by reading the remaining three answer choices. Instead, formulate a new prediction and find a match among the three remaining answer choices.

> **MCAT Expertise**
>
> If you read a question stem and it doesn't give you very much to work with, don't just say *I don't know* and jump straight to the answer choices. Use your Outline to remember the main themes of the passage and then use those themes to help with process of elimination. This will help you avoid being distracted by answer choices that are seductive but that do not fit into the passage.

Sometimes the question stem just doesn't give you very much to work with, and on other occasions you'll search through the answers but find no likely match. In these cases, you will have to use process of elimination, which may require multiple returns to the text as you research each choice individually. If you were able to set expectations during the Execute step for wrong choices, however, less additional research will be required. When evaluating choices, remember that the answer choices will sometimes vary the language used in the passage. Don't rule out a choice just because the wording is not quite what you anticipated; consistency of concepts matters more than identical terminology. Keep in mind that an answer requires only one major flaw for elimination, so the Wrong Answer Pathologies described later in this chapter can greatly expedite the process.

When all else fails, you can fall back on educated guessing. Eliminate whatever you can and then go with your gut among the remaining options. Never make a blind guess unless you're completely out of time and need to fill in an answer choice. Even crossing off just one wrong answer will increase your chances of randomly choosing the correct one by 33 percent. Crossing off two wrong answers will double your chances. If possible, work on any unanswered questions for the passage and see if that effort allows you to return to rule out additional incorrect options.

> **MCAT Expertise**
>
> **Should I compare answer choices?** Your default assumption should be that only one answer choice is flawless and that the others contain at least one flaw each, sufficient for ruling them out. However, you may occasionally find questions containing superlatives (*strongest challenge*, *most supported*, *best example*, and so on) in which you need to compare two or more answers that have the same effect but to different degrees. When making such comparisons, don't assume that an Extreme answer is necessarily wrong, especially if the question stem includes the words *if true* or similar language. A stronger answer that nevertheless produces the desired outcome would be the correct choice.

8.2 Wrong Answer Pathologies

The AAMC has designed the CARS section to be a fair test of critical thinking skills. The need for fairness is great news because it means that the questions won't play tricks on you! There will never be a question with two correct answer choices or one in which all of the options are wrong. Each question you encounter on Test Day will have one and only one right answer and three that are incorrect for at least one reason. Even better, there can only be so many of these reasons: in fact, a few of them are found so frequently that you can treat them like recurring signs and symptoms of answer choice "illness." Naturally, we call them **Wrong Answer Pathologies**.

A choice only needs one fatal flaw to be worth eliminating, but often wrong answer options have many issues, so don't necessarily be alarmed if you ruled out a wrong answer for a different reason than the one mentioned in a practice question's explanation. In addition to having some occasional overlap, the following list of pathologies is not meant to be exhaustive; it includes only the four patterns we've identified as the most common through researching all of the released MCAT material. In the Kaplan Method for CARS Questions just detailed, pathologies function as recurring expectations for wrong answers, which you can assume fit for most of the questions you encounter (with a few significant departures noted below).

Note: The Wrong Answer Pathologies, as well as the Kaplan Method for CARS Passages, Kaplan Method for CARS Questions, and CARS Question Types are included as tear-out sheets in the back of this book.

FAULTY USE OF DETAIL (FUD)

The testmakers will often include accurate reflections of passage details in wrong answers, primarily to appeal to those students who jump at the familiar language. What makes the use of a detail "faulty" is that it simply doesn't answer the question posed. It may be too specific for a question that requires a general answer, or it may be that the detail comes from the wrong part of the passage. Even if a choice comes from the right paragraph, the detail cited might not be relevant to the question posed, which is often the case in Strengthen–Weaken (Within the Passage) questions. A thorough prediction makes catching these FUDs much easier.

OUT OF SCOPE (OS)

With the noteworthy exception of *Reasoning Beyond the Text* questions (for which this pathology does not apply), an answer choice that is outside the scope of the passage will inevitably be wrong. Typically, such answers will be on topic but will bring in some element that the text does not discuss. For instance, if an author never makes comparisons when discussing different ideas, an Out of Scope answer choice might involve the author ranking two or more concepts. Another common OS pattern is the suggestion that a view was the first of its kind or the most influential, when the author entirely avoids discussing its historical origins or relative popularity or influence. Keep in mind that information can be unstated by the passage but not count as

---

Key Concept

Wrong Answer Pathologies are the most frequent patterns found in incorrect answer choices. They are so common that you'll find at least one in just about every CARS question and even in many of the questions in the three science sections!

Real World

A Faulty Use of Detail answer choice is like a politician who dodges a question during a debate:
- Moderator: *Candidate X, what will you do to improve the economy?*
- Candidate X: *You know, that's a wonderful question. My platform is to stand for all Americans' rights and to represent them fairly. If elected, I will help lead our country with integrity.*

True as the claim may be, the response just doesn't answer the question posed!

Bridge

The answers to many *Reasoning Within the Text* questions will be claims unstated by the passage that can nonetheless be inferred. Remember that inferences include assumptions (unstated evidence) and implications (unstated conclusions). Inferences are discussed in Chapter 5 of *MCAT CARS Review*.

MCAT Critical Analysis and Reasoning Skills

Out of Scope, as will be the case with the correct answers to many *Reasoning Within the Text* questions, so don't be too quick to reject a choice as OS just because the author does not explicitly say it.

OPPOSITE

Whenever an answer choice contains information that directly conflicts with the passage, we call it an Opposite. Often the difference is due simply to the presence (or absence) of a single word like *not* or *except*, a prefix like *un–* or *a–*, or even a suffix like *–less* or *–free*. Be especially careful when stems or choices involve double (or triple) negatives; they're much less difficult to Assess if you reword them with fewer negations. Moreover, don't assume that just because two answer choices contradict each other that one of them has to be correct. For example, suppose an author argues that it is impossible to prove whether or not a divine being exists, a variant of the religious view known as agnosticism. If a question accompanying the passage were to ask for a claim the author agreed with, *God exists* and *There is no God* would both be Opposites of the correct answer.

DISTORTION

Extreme answers and others that "twist" the ideas in the passage further than the author would prefer are what we call Distortions. Although they do not automatically make a choice incorrect, the following are common signals of distorted claims:

- Strong words like *all*, *always*, *none*, *never*, *impossible*, or *only*
- A prefix like *any–* or *every–*
- A suffix like *–est* or *–less*

MCAT authors typically do not take radical positions on issues, so it's worth noting whenever they do. In those rare cases, Extreme choices would not actually be Distortions of the author's view and might be correct. The other major case in which Extreme answer choices should not be immediately ruled out is when the question stem tells you that you can treat the answer choices as true and your task is only to gauge which would have the greatest impact on a particular argument. This is often the case with Strengthen–Weaken (Beyond the Passage) questions.

8.3 Signs of a Healthy Answer

If you're like most students prepping for the CARS section, you've had a dispute with at least one question explanation. *Hey, what about what the author says in the first paragraph?* you may have wondered, or perhaps you've said to yourself (or aloud!) *But couldn't you think of it like this instead?* While you may be in the habit of arguing for points with college professors, it does you no good to try to argue with the MCAT. The testmakers are extremely deliberate about how they word correct answers, always taking care to include exactly one per question.

Correct answer choices can vary widely in appearance, but there are patterns in how they are written as well. If the traps that can lead you astray on Test Day are appropriately called Wrong Answer Pathologies, then these corresponding traits can be thought of as indicators of good health. While the following signs are not enough by themselves to make an answer right, you can generally expect them to correspond to the correct choices in most types of questions.

APPROPRIATE SCOPE

You might say correct answers follow the "Goldilocks principle" when it comes to scope: not too broad, not too specific, but just right. The **scope** defines the limits of the discussion, or the particular aspects of the larger topic that the author really cares about. In your Outline, the Goal that you jotted down should give you an idea of the scope of the passage overall. As a general rule (with one important exception), correct answers to MCAT questions will remain within the scope of the passage, but you can formulate a more precise expectation of what scope the correct answer needs to have by identifying the question's type and task.

Main Idea questions will always have correct answers that match the scope of the entire passage. They will typically include at least one wrong answer that is too focused (Faulty Use of Detail) and at least one that goes outside the passage entirely (Out of Scope). In contrast, Detail and Definition-in-Context questions usually require more refined scopes to identify their correct answer choices. If a clue directs you to a particular portion of the passage, the correct answer, more often than not, will have the same scope as the referenced text (or what immediately surrounds it).

The important exception to the rule that answers must remain within the scope of the author's discussion applies to the category of *Reasoning Beyond the Text* questions, addressed in Chapter 11 of *MCAT CARS Review*. As their name suggests, these broaden the scope to new contexts, sometimes appearing to have no connection to the passage whatsoever. Note, however, that some *Reasoning Beyond the Text* questions will present new information in the stem but have answers that stick to the scope of the passage anyway. So be savvy with the answer choices in *Reasoning Beyond the Text* questions: while the correct answer choice will tend to lie slightly outside the scope of the passage, don't automatically rule out an answer choice just because it *happens* to be in scope.

> **MCAT Expertise**
>
> The scope of a text refers to the particular aspects of a topic that the author addresses. Every paragraph in a CARS passage has its own scope, and together you can think of them as constituting the scope of the whole passage. Similarly, each answer choice will have its own scope, which could mimic any part of the author's discussion or depart from the passage entirely. It is essential to note that having the same scope doesn't necessarily mean having identical content. For instance, unstated assumptions in an argument are definitely within the scope of the passage, even though the information they contain is left unsaid by the author.

AUTHOR AGREEMENT

Unless a question stem explicitly asks about an alternative viewpoint or a challenge to the information presented in the passage, a correct answer choice will be consistent with what the author says. This is one reason why considerations of *tone* and *voice* (most clearly reflected by Author keywords) are usually important enough to be worth including in your Passage Outline, as was recommended in Chapter 4 of *MCAT CARS Review*. Generally, a correct answer should not contradict anything that the author says elsewhere in the passage, with the possible exception of sentences that speak in a different voice than the author's (such as quotes or references to others' opinions). In short, if it doesn't sound like something the author would say, you'll most likely want to rule it out.

WEAKER IS USUALLY BETTER

One final consideration is a consequence of the fact that the AAMC tends to select passages by authors who do not take extreme views. You may find one or two passages on Test Day with more radical writers; for them, a stronger claim in the answer choices may actually be a good sign. However, for most of the passages you'll encounter, authors tend to use numerous Moderating keywords to limit the strength of their claims. Because a stronger claim has a higher burden of proof (that is, stronger evidence must be provided to support the claim), most authors avoid them to make what they write seem more plausible. Thus, you should generally give preference to answer choices that use weaker language such as *can*, *could*, *may*, *might*, *is possible*, *sometimes*, *often*, *likely*, *probably*, and *in some sense*. Exceptions to this tendency were addressed earlier in the discussion of Distortion.

Conclusion

This chapter is only an introduction to the question method; the three chapters that follow are a necessary supplement for seeing how the method functions in practice. Specific strategy suggestions and worked examples are included for each of the most common question types and tasks, together constituting well more than 90 percent of what you'll encounter on Test Day. The explanations accompanying these sample questions will also identify their Wrong Answer Pathologies, giving you some concrete examples to go with the explanations provided here.

CONCEPT AND STRATEGY SUMMARY

Kaplan Method for CARS Questions
- **Assess**
    - Read the question, **NOT** the answers
    - Identify the question type and difficulty
    - Decide to attack *now* or *later*
- **Plan**
    - Establish the task set by the question type
    - Find clues in the stem on where to research
    - Navigate the passage using your Outline
- **Execute**
    - Predict what you can about the answer
    - Set expectations for wrong choices
    - Be flexible if your first Plan flops
- **Answer**
    - Find a match for your prediction, or
    - Eliminate the three wrong options, or
    - Make an educated guess

Wrong Answer Pathologies
- **Faulty Use of Detail (FUD)** answer choices may be accurate statements, but they fail to answer the question posed.
    - The answer choice may be too specific for a question that requires a general answer.
    - The answer choice may use a detail from the wrong part of the passage.
    - The answer choice may be from the right paragraph but still not be relevant to the question posed.
- **Out of Scope (OS)** answer choices usually bring in some element that the passage does not discuss (and that cannot be inferred from the passage).
    - The answer choice may make connections or comparisons that the author did not discuss.
    - The answer choice may make a statement about the significance or history of an idea that the author did not.
    - The answer choice may otherwise bring in information that does not fall within the constraints of the passage.

- **Opposite** answer choices contain information that directly conflicts with the passage.
  - The answer choice may contain (or omit) a single word like *not* or *except*.
  - The answer choice may contain a prefix like *un–* or *a–* or a suffix like *–less* or *–free*.
  - The answer choice may say that a given claim is true, when the author is ambivalent.
- **Distortion** answer choices are extreme or twist the ideas in the passage further than the author would prefer.
  - The answer choice may use a strong word like *all, always, none, never, impossible*, or *only*.
  - The answer choice may contain a prefix like *any–* or *every–* or a suffix like *–est* or *–less*.
  - The answer choice is usually more radical than the author because radical positions are hard to support and are rare in MCAT passages.

Signs of a Healthy Answer

- Correct answers tend to have the right **scope**—not too broad, not too specific, but just right.
- Correct answers tend to be consistent with the author's statements and opinions.
- Correct answers tend to use Moderating keywords, such as *can, could, may, might, is possible, sometimes, often, likely, probably*, and *in some sense*.

8: Question and Answer Strategy

WORKED EXAMPLE

Use the Worked Example below, in tandem with the subsequent practice passages, to internalize and apply the strategies described in this chapter. The Worked Example matches the specifications and style of a typical MCAT *Critical Analysis and Reasoning Skills* (CARS) passage.

Passage	Analysis
Mayan signs are by nature highly pictorial, often representing in considerable detail animals, people, body parts, and objects of daily life. The pictorial principle is taken to the extreme in inscriptions composed of "full-figure" glyphs, in which individual signs and numbers become animated and are shown interacting with one another. None of this should be taken to mean that the Mayans had simple picture writing. The Mayans wrote both logographically and phonetically, and within its phonetic system alone, the Mayans had multiple options. All English words are formed from various combinations of only 26 phonetic signs. By contrast, all Mayan words can be formed from various combinations of nearly 800 consonant–vowel glyphs, each representing a full syllable. Sounds are formed by combining a particular consonant with one of the five vowels (hence a syllabary, rather than alphabet).	Glancing at the first paragraph, we see an introduction to Mayan writing. The author specifically picked out two defining characteristics: logographic and phonetic—we should expect to see more about these in the rest of the passage. We should not worry about the definition of *logographic*; as we continue through the passage, we get the gist as "pictorial." The author finishes by contrasting the complexity of Mayan writing with English. The Label could be: **P1.** Mayans wrote in pictures but their language was complex
The combination of consonant–vowel syllabic glyphs and logographs enabled the scribes a variety of choices to write the words of their texts in detail. For example, one very common honorific title in Mayan texts is *ahaw*, meaning "lord" or "noble." Ahaw may be written in logographic form as a head in profile, with the distinctive headband or scarf that marked the highest nobility in Mayan society. But it is also possible to write the word as a combination of three phonetic, syllabic signs: *a–ha–wa*. Likewise, the word *pakal* ("shield") can be indicated by a depiction of a shield or by the combination of syllabic elements *pa–ka–la*.	In this paragraph, we see the complexity that the author discussed earlier with two examples. The simple word *ahaw* can be written or drawn in several ways—either as a single *logographic form* or as multiple *phonetic, syllabic signs*. The second example, *pakal*, works similarly. A Label of this paragraph could be: **P2.** Logograph (picture) or multiple syllables can be used for same word

Passage	Analysis
Because many Mayan signs remain undeciphered, it's not possible to state precisely the relative proportions of logographic and syllabic signs. But a significant number of the logograms have been deciphered and the number of deciphered syllabic signs keeps growing. Epigraphers have filled more than half of the syllabic grid (which plots the consonants of the spoken Mayan language against its vowels and thus represents the totality of signs needed to write the language). Half of the grid may seem a meager proportion, but it must be remembered that the discovery of the structure of the syllabic elements—Knorozov's main contribution—was made only a little more than 30 years ago. Furthermore, the consonant–vowel syllables that are already understood are the common ones. Many of the empty spaces in the syllabic grid remain so because they are linguistically rare; rare signs are more difficult to translate than common ones.	This paragraph starts right off by telling us that *many Mayan signs remain undeciphered*, but much has been learned about the language. According to the author, *a significant number of the logograms have been deciphered*, but only about half of the syllabic signs. This is justified with two claims: first, that some person *Knorozov* discovered the *structure of the syllabic elements* only 30 years ago; and second, that the syllables that we know are *common*—the rest are rare, which makes them *more difficult to translate*. The gist of this paragraph, and therefore a good Label, is: **P3.** Describes progress in deciphering the syllabic grid
Nonetheless, the pace of phonetic decipherment is bound to increase in the coming years as more resources are trained on it. One aspect of Mayan writing that may complicate this progress is the fact that different signs can have the same value. Two signs that share a value are known as allographs. Such equivalences are common in Mayan texts (there are at least five different signs that could be chosen to represent the Mayan syllable *ba*). Each scribe chose from several different signs to convey the sounds. In evaluating a particular phonetic interpretation of a syllable, it's helpful to identify as many as possible of the variant forms; so the process of recognizing allographs depends on the slow work of comparing many texts in order to find variant spellings of the same word.	We see *Nonetheless*, which immediately indicates a Difference from the previous idea. The author tells us that the rate of *phonetic decipherment is bound to increase*. The reason that deciphering has been slow is *allographs*, or different symbols that represent the same entity. The Label for this final paragraph could be: **P4.** Decipherment progress should increase: complication = allographs

8: Question and Answer Strategy

Here's a sample Outline and Goal for this passage:

P1. Mayans wrote in pictures but their language was complex

P2. Logograph (picture) or multiple syllables can be used for same word

P3. Describes progress in deciphering the syllabic grid

P4. Decipherment progress should increase: complication = allographs

Goal: To explain Mayan writing forms and suggest that decipherment of the syllabic grid will increase

Question	Analysis
1. The author mentions Knorozov in the third paragraph in order to:	This is a Function question, which is a *Foundations of Comprehension* question. Because we read the passage for perspective, this is a question that we can do right away. Where did we see *Knorozov*? Paragraph 3 deals with progress in completing the syllabic grid. *Knorozov* was used to justify why only about half of the syllabic grid has been filled in—he discovered the syllable structure only *a little more than 30 years ago*.
A. prove that the recent discovery of Mayan signs has led to its lack of decipherment.	**(A)** is a Distortion on two counts: first, the recent discovery is of the syllable structure of the language, not the *Mayan signs* themselves; second, *prove* is far too strong a word—the author is not nearly so Extreme.
B. offer an explanation for what may appear to be a relative paucity in the completion of the Mayan sign syllabic grid.	**(B)** fits and is the correct answer. Knorozov and his recent discovery is part of the justification for why there is so little filled in on the syllabic grid.
C. argue that expert linguists have been stymied in their attempts to decipher and understand many allographic Mayan signs.	**(C)** is a Faulty Use of Detail; the challenge presented by allographs is described in paragraph 4 and is not associated with Knorozov.
D. show how the understanding of other linguistic structures may improve the comprehension of Mayan syllabic signs.	**(D)** is a classic Out of Scope answer choice as *other linguistic structures* are never described in the text.

Question	Analysis
2. As used in the passage, the term "logographic" most closely refers to:	This is a Definition-in-Context question. As another *Foundations of Comprehension* question, we can do this right away. We saw *logographic* in paragraphs 1 and 2. When introduced in paragraph 1, the gist of the word was that it was a picture. Paragraph 2 then describes the writing of the word *Ahaw*: *Ahaw may be written in logographic form as a head in profile*. Because a picture is used to represent the word in *logographic form*, they must be pictorial symbols.
A. a written phonetic representation of a word. B. syllabic division of an individual word.	**(A)** and **(B)** are both Opposites that describe the other form of Mayan writing presented in the passage.
C. imagistic representation of an idea.	**(C)** simply paraphrases the idea of pictorial representation, so it's the correct answer.
D. a visual picture of an idiomatic phrase.	**(D)** is a Distortion; although logographs are visual representations, the author suggests that they represent single words, not entire *idiomatic phrase[s]*.

8: Question and Answer Strategy

Question	Analysis
3. The author of the passage would be LEAST likely to agree with which of the following?	This is an Apply question. Because this is a *Reasoning Beyond the Text* question type, this is one that we should save for later. The question asks what *the author...would be LEAST likely to agree with*, so the answer should be inconsistent with the author's beliefs. The author believes that Mayan writing is complex for several reasons but is steadily being deciphered. We want an answer choice that goes directly against this premise.
A. Languages with writing composed of pictorial signs can demonstrate a remarkable degree of complexity and detail.	**(A)** is an Opposite; the author would certainly agree with this statement given his or her belief that the language is more than just *simple picture writing*; the description of the *800...glyphs*, multiple ways of writing the same word, and allographs support this idea.
B. Linguistic signs based on syllabic or phonetic coding may be easier to decipher than those based on visual images.	**(B)** is Out of Scope; this comparison is never made, so it's impossible to say whether the author would agree with this statement or not.
C. Pictorial languages are restricted to the expression of simple ideas because of their emphasis on image.	The author would certainly disagree with **(C)**, making it the correct answer. Mayan writing doesn't convey only *simple ideas*, the author would argue, because it's more than just *simple picture writing*.
D. The existence of allographs in Mayan signs indicates the complexity of this linguistic system.	**(D)** is an Opposite; this summarizes the point of paragraph 4: allographs are a complication that makes deciphering Mayan writing more difficult.

MCAT Critical Analysis and Reasoning Skills

Question	Analysis
4. The author discusses the words *ahaw* and *pakal* in order to:	This is another Function question, so we should do it right away. Where did we see the words *ahaw* and *pakal*? The words were in paragraph 2. The purpose of paragraph 2 is to explain that a logograph (picture) or multiple syllables can be used to represent the same word.
A. estimate the number of meanings that some common Mayan words may possess.	(A) Distorts the author's point: the same word can have multiple ways of being represented, not multiple *meanings*.
B. compare the flexibility of Mayan logographs to that of consonant–vowel syllables.	(B) is Out of Scope as the author does not suggest a comparison between the *flexibility* of phonetic and logographic writing—just that they can both be used to write certain words.
C. illustrate the difficulty of understanding detailed Mayan texts.	(C) is a Faulty Use of Detail; even *though detailed Mayan texts* are difficult to understand, this is a point from paragraph 4 rather than paragraph 2, where *ahaw* and *pakal* are introduced.
D. demonstrate that Mayan words may appear in both logographic and syllabic form.	(D) matches cleanly with the prediction and the Label for paragraph 2 and is the correct answer.

8: Question and Answer Strategy

Question	Analysis
5. According to the author, which of the following might best address some of the decipherment problems associated with Mayan signs?	This is an Inference question, asking for a conclusion the author did not explicitly state, or an Implication. Because the example is not explicitly stated in the passage, we should save this question until later. Problems with *decipherment* are mentioned in paragraphs 3 and 4. In paragraph 3, the author discussed that the remaining unknown syllables are rarely used and thus harder to translate. Paragraph 4 continued the idea by talking about allographs. The author does say in the last lines of the passage that deciphering allographs depends on *comparing many texts*. Look for an answer that sounds like the points in either of these two paragraphs.
A. Additional financial and scholarly resources should be directed towards this linguistic effort.	**(A)** is a Distortion; the author argues at the beginning of paragraph 4 that the *pace of phonetic decipherment is bound to increase* as more resources are trained on it, but stops short of saying that more resources *should* be trained on it, or suggesting what *form* those resources might take. This answer choice changes a statement of fact to a recommendation as well as interpreting *resources* far more specifically than can be justified by the passage.
B. More attention should be focused on identifying logographic signs than on categorizing syllabic signs.	**(B)** is an Opposite; because allographs are phonetic representations, the author would argue that if anything, there should be more attention given to syllabic signs than logographic signs—many of which have already been deciphered.
C. Scholars should prioritize the completion of the syllabic grid.	**(C)** Distorts the author and is too Extreme; the author actually seems pretty content with the current, incomplete state of the grid and its slow but steady progress toward being filled in.
D. Careful study of comparative texts should continue in order to evaluate phonetic interpretation of each syllable.	**(D)** matches closely statements at the end of paragraph 4 that *the process of recognizing allographs depends on the slow work of comparing many texts in order to find variant spellings of the same word.*

Question	Analysis
6. The author implies which of the following about the ratio of logographic to syllabic signs in Mayan writing?	This is an Inference question, asking us to *Reason Within the Text*. We should save this question for later. The beginning of paragraph 3 discussed the ratios between the two types of signs. The author says that *it's not possible to state precisely the relative proportions of logographic and syllabic signs* because many of the signs are undeciphered. The clear implication is that as these signs are deciphered, the ratio will become clearer.
A. Researchers disagree about the correct way to determine it.	**(A)** is Out of Scope; the author suggests that the lack of a ratio is due to a lack of information, not because of *disagree[ment]*.
B. Its practical value has failed to attract serious attention.	**(B)** is also Out of Scope as the author doesn't indicate anything about the *practical value* of knowing this ratio.
C. A meaningful ratio may never be established.	**(C)** is a Distortion; the author mentions that the number of deciphered signs is growing. So, if anything, the author would argue that a ratio may be established some day in the future.
D. More work must be done before the ratio can be determined.	**(D)** matches perfectly. Once all logographs and syllables are translated, we would know the ratio—but that will require *more work*.

8: Question and Answer Strategy

Question	Analysis
7. Which of the following would most call into question the author's argument about the complexity of Mayan writing?	This is a Strengthen–Weaken (Beyond the Passage) question, so we should save it until last. We need to *call* the *complexity of Mayan writing into question*; this means we need to find that answer that would make the author's conclusion less likely. Where do we see the author's argument about the complexity of Mayan writing? In paragraph 3, the author states that Mayan writing is complex due to the fact that it contains both logographic and syllabic writing that can be interchangeable. In paragraph 4, the author goes on to state that part of the problem in deciphering Mayan writing is the presence of so many allographs that scribes had to choose from when writing. To make these ideas less likely, we need to find a statement that makes the combination of logographs and symbols easy to decipher or tells us that allographs did not exist.
A. It has recently been discovered that allographs are actually just stylistic differences between scribes.	**(A)** is exactly what we are looking for: if allographs are actually just stylistic differences—and not *different signs that could be chosen to represent* the same syllable, then the language would likely become easier to translate.
B. The syllabic grid is only partially complete.	**(B)** is a statement made by the author that has no strong relationship with the *complexity of Mayan writing*—only a relationship with the process of translating the writing. This statement has no effect on the author's argument besides being mentioned in the passage.
C. Other pictorial languages, such as Egyptian, have been deciphered. D. Languages based on logographs are less complicated than modern syllabic languages.	**(C)** and **(D)** are Out of Scope as the author is not concerned with other languages outside of Mayan.

Practice Questions

Passage 1 (Questions 1–7)

The rich analysis of Fernand Braudel and his fellow *Annales* historians have made significant contributions to historical theory and research, not the least of which is a broad expansion of potential routes of scholastic analysis. In a departure from traditional approaches, the *Annales* historians assume that history cannot be limited to a simple recounting of conscious human actions, but must be understood in the context of forces and material conditions that underlie human behavior. Braudel was the first *Annales* historian to gain widespread support of the idea that history should synthesize data from various social sciences, especially economics, in order to provide a broader view of human societies over time (although Febvre and Bloch, founders of the *Annales* school, originated this approach).

Braudel conceived of history as the dynamic interaction of three temporalities. The first of these, the *evenementielle*, involved short-lived dramatic "events," such as battles, revolutions, and the actions of great men, which had preoccupied traditional historians like Carlyle. *Conjonctures* was Braudel's term for larger cyclical processes that might last up to half a century. The *longue durée*, a historical wave of great length, was for Braudel the most fascinating of the three temporalities. Here he focused on those aspects of everyday life that might remain relatively unchanged for centuries. What people ate, what they wore, their means and routes of travel—for Braudel, these things create "structures" that define the limits of potential social change for hundreds of years at a time.

Braudel's concept of the *longue durée* extended the perspective of historical physical space as well as time. Until the *Annales* school, historians had generally taken the juridical political unit—the nation-state, duchy, or whatever—as their starting point. Yet, when such enormous timespans are considered, geographical features may well have more significance for human populations than national borders. In his doctoral thesis, a seminal work on the Mediterranean during the reign of Philip II, Braudel treated the geohistory of the entire region as a "structure" that had exerted myriad influences on human lifeways since the first settlements on the shores of the Mediterranean Sea. And so the reader is given such esoteric information as the list of products that came to Spanish shores from North Africa, the seasonal routes followed by Mediterranean sheep and their shepherds, and the cities where the best ship timber could be bought.

Braudel has been faulted for the impression of his approach. With his Rabelaisian delight in concrete detail, Braudel vastly extended the realm of relevant phenomena; but this very achievement made it difficult to delimit the boundaries of observation, a task necessary to beginning any social analysis. Indeed, to identify an appropriate scope of work when all options and directions for academic inquiry are available before one's eyes is a nearly impossible task. Further, Braudel and other *Annales* historians minimize the differences among the social sciences. Nevertheless, the many similarly designed studies aimed at both professional and popular audiences indicate that Braudel asked significant questions that traditional historians had overlooked.

1. The author refers to the work of Febvre and Bloch in order to:
   A. illustrate the need to delimit the boundaries of observation.
   B. suggest the relevance of economics to historical investigation.
   C. debate the need for combining various social science approaches.
   D. show that previous *Annales* historians anticipated Braudel's focus on economics.

2. In the third paragraph, the author is primarily concerned with discussing:
   A. Braudel's fascination with obscure facts.
   B. Braudel's depiction of the role of geography in human history.
   C. the geography of the Mediterranean region.
   D. the irrelevance of national borders.

3. The passage suggests that, compared with traditional historians, *Annales* historians are:
   A. more interested in other social sciences than in history.
   B. critical of the achievements of famous historical figures.
   C. skeptical of the validity of most economic research.
   D. more interested in the underlying context of human behavior.

4. Which of the following statements would be most likely to follow the last sentence of the passage?
   A. Few such studies, however, have been written by trained economists.
   B. It is time, perhaps, for a revival of the Carlylean emphasis on personalities.
   C. Many historians believe that Braudel's conception of three distinct "temporalities" is an oversimplification.
   D. Such diverse works as Gascon's study of the Lyon and Barbara Tuchman's *A Distant Mirror* testify to his relevance.

5. Some historians are critical of Braudel's perspective for which of the following reasons?
   A. It seeks structures that underlie all forms of social activity.
   B. It assumes a greater similarity among the social sciences than actually exists.
   C. It fails to consider the relationship between short-term events and long-term social activity.
   D. It rigidly defines boundaries for social analysis.

6. Which of the following historical phenomena would the author most likely consider an example of the *longue durée*?
   A. Julius Caesar's crossing of the Rubicon, which led to a four-year civil war in Rome
   B. The occurrence in ancient Rome of devastating malaria outbreaks roughly every half-century
   C. A gradual shift towards a drier Mediterranean climate that lasted from 250 to 600 A.D. and impacted food availability and daily life
   D. The battle of Carrhae in 53 B.C.E., in which the Parthians exterminated a large Roman army

7. Suppose Braudel was once quoted as saying, "For me, the most interesting aspects of history are those in which one man's decisions turned the tides of history." What relevance would this new information have for the passage?
   A. It would weaken the author's claim that Braudel was most interested in the *longue durée* temporality.
   B. It would weaken the author's claim that Braudel was not interested in the actions of great men.
   C. It would strengthen the author's claim that Braudel was most interested in the *evenementielle* temporality.
   D. It would strengthen the author's claim that Braudel considered the best historical analysis to focus on the actions of great men.

Passage 2 (Questions 8–12)

A balance of power arises when a group of neighboring countries enters a state of economic and military equilibrium. In a balance of power system, a nation-state cannot violate the independence or the essential rights of another without incurring reprisals from neighboring states acting to restore balance. Conflict ought to be avoided, but may be necessary when a potential hegemon takes aggressive action. This is because peace or the safety of an individual nation is less important in balance of power than preserving equilibrium in the system. Diplomacy and trade in a balance of power is thus a continuum of action and reaction, rather than a series of attempts at independent policy-making.

What conditions are necessary for a balance of power to occur? First, states must be aligned in a "state system." The states involved must be independent, close in proximity (often possessing shared borders), and near equals in power. When one state far outstrips its close neighbors in power, it dictates economic and military policy for the region. When neighbor states are commensurate, however, interaction on issues of shared concern occurs on a level playing field of policy-making.

A second factor in the formation of a balance of power is the framework of the system itself. To maintain an effective balance of power, a system must include a minimum of three states. A true equilibrium cannot exist between two states because one state inevitably gains ascendancy over the other. A balance of power is also generally characterized by a common ground of culture in the state system. A group of states is more likely to align in a cooperative manner if constituent states perceive a degree of cultural similarity with their neighbors. Added to this, the mechanism of diplomacy, skilled diplomats, and economic alliance structures must be in place for a balance of power to thrive. If this framework exists, then the system will be sufficiently flexible to survive short-lived economic fluctuations and military aggression.

Inherent in this need for a framework is the third precondition for the evolution of a balance of power: rational estimation. Countries involved in a balance of power must have a rational means of estimating the power of individual and combined states within the system. The flow of economic information between countries is crucial for an economic equilibrium to persist; a nation's economic planning should consider the dynamics of the entire state system in addition to its own agriculture and industry. A similar diffusion of information must occur on national security issues. This means that the stability of a state system depends on the development of sophisticated intelligence agencies for estimating the military capability and activity of other states.

The importance of military intelligence is perhaps best illustrated on a smaller scale. Suppose that your neighbor has stolen your lawnmower and you are considering taking retribution by picking a fight with him. For your survival and for the continued survival of the balance of power in your neighborhood, you must first be aware of all relevant personal defense issues. In particular, it is critical that you find out whether your neighbor possesses a gun and whether he might use it under such circumstances. Without the gathering of such information, the balance of power in the neighborhood cannot be maintained.

8. Which of the following helps explain why a balance of power is generally characterized by a "common ground of culture" (paragraph 3)?

   A. Countries with similar cultures often have mechanisms of diplomacy in place.
   B. Cultural differences between two nations are often the source of military conflict.
   C. A hegemon often influences the cultures of all nations surrounding it.
   D. Alliances among nations are more common if there are shared cultural beliefs.

9. The author's use of the phrase "level playing field" (paragraph 2) is probably meant to indicate that in a balance of power:

   A. states coordinate policy in a democratic manner.
   B. diplomatic initiatives generally work to the advantage of all member states.
   C. each state has some influence on economic and military events.
   D. member states form alliances against countries outside the balance of power.

10. The passage suggests that in order to participate in a balance of power, a country should:

   I. plan its agricultural and industrial development.
   II. possess information on its neighbors' economies.
   III. disclose its military secrets to its neighbors.

   A. II only
   B. I and II only
   C. I and III only
   D. I, II and III

11. The analogy between an international balance of power and the interrelations between neighbors is based on which of the following assumptions?

   A. Both neighbors and neighbor states must avoid conflict whenever possible.
   B. Power relations between neighbors are dependent on similar factors to those between neighbor states.
   C. A degree of cultural similarity is required for peaceful coexistence.
   D. The use of force for self-preservation is always justifiable.

12. According to the passage, which of the following would be LEAST critical to the survival of a neighborhood's balance of power?

   A. An alliance network of a minimum of five or more neighbors
   B. The presence of a number of skilled negotiators
   C. Information about the location of dangerous weapons
   D. Knowledge of neighbors' typical behavior in times of conflict

Explanations to Practice Questions

Passage 1 (Questions 1–7)

Sample Passage Outline

P1. *Annales vs.* traditional historians; Braudel popularized

P2. Braudel's three temporalities

P3. Expansion on *longue durée* as geohistory

P4. Braudel critics: "scope too broad, differences of social sciences blurred"; Author: "but asked new questions"

Goal: To explain Braudel's impact on historical theory

1. D

A classic Function question: Why did the author refer to Febvre and Bloch? Go back to the end of paragraph 1 where Febvre and Bloch are mentioned and read the surrounding text: *Braudel was the first* Annales *historian to gain widespread support of the idea that history should synthesize data from various social sciences…(although Febvre and Bloch, founders of the Annales school, originated this approach)*. In other words, while Braudel popularized this approach, it was actually Febvre and Bloch who came up with it in the first place. This prediction matches closely with **(D)**, which has Febvre and Bloch *anticipat[ing]* Braudel's approach. **(A)** and **(C)** both suggest that Febvre and Bloch could be used to argue against the *Annales* approach, which is inconsistent with their roles as originators of this same approach. **(B)** tries to Distort the author's use of Febvre and Bloch: economics is clearly relevant for the *Annales* approach to historical investigation, but Febvre and Bloch are not used in the context of supporting the relevance of economics—just that they used economics before Braudel.

2. B

This is another Function question that essentially asks, *What is the role of the third paragraph*? The Label for paragraph 3 gives us the answer: *Expansion on* longue durée *as geohistory*. This prediction matches closely with **(B)**, which reflects on *the role of geography in human history*. **(A)** is a Faulty Use of Detail wrong answer; the author does mention Braudel's attention to *esoteric information*, but this is a minor detail from the paragraph that misses the full purpose of expanding on the *longue durée*. **(C)** is also a Faulty Use of Detail; the *geography of the Mediterranean region* is used to support the use of geography in Braudel's approach, but the primary concern is not to discuss *the geography of the Mediterranean* specifically—rather, it is to show how geography can influence human history. **(D)** Distorts the author's words. Braudel does shy away from national borders to focus on geographic borders, but that does not mean that all national borders are *irrelevan[t]*.

3. D

This is an Inference question. Based on the evidence presented in the passage, we should be able to infer a difference between traditional and *Annales* historians. From our Outline, we know that the two are contrasted in paragraph 1, so this is where we will look for clues. The key to the answer is given in the second sentence: *In a departure from traditional approaches, the* Annales *historians assume that history…must be understood in the context of forces and material conditions that underlie human behavior*. In other words, the *Annales* historians are *more interested in the underlying context of human behavior* than traditional historians, as **(D)** states. **(A)** Distorts the author's words: the *Annales* are interested in incorporating social sciences into historical analysis, but that does not mean that they are more interested in other social sciences than in history—they are historians after all! **(B)** is also a Distortion. The *Annales* historians propose that history is more than just the actions of famous figures, but this doesn't mean that they are *critical of* those figure's *achievements*. Finally, **(C)** is an Opposite. *Annales* historians want to incorporate economic research findings into historical analysis and so should not be *skeptical of the validity* of such approaches.

4. D

We want to predict what direction the author would go in if the passage were continued in this *Reasoning Beyond the Text* question. Recall that the purpose of the last paragraph was to mention criticisms of Braudel's approach and respond to those criticisms. The paragraph ends by responding to criticism by citing the contribution of Braudel's work: *studies…indicate that Braudel asked significant questions that traditional historians had overlooked*. If another sentence were added, it should continue along the same lines of highlighting the influence or merits of Braudel. This prediction matches perfectly with **(D)**. **(A)** would backtrack on the author's support by questioning the professionalism of the authors of such studies. **(B)** is both Out of Scope and Opposite. Incorporation of the *Carlylean approach* mentioned in paragraph 2 would come out of nowhere at the end of paragraph 4, and it also represents the traditional approach to historical analysis; the author is unlikely to support a revival of this approach. Finally, **(C)** is a criticism of Braudel, and because the author is defending Braudel in the last part of the paragraph, this answer choice is an Opposite.

5. B

This is a Detail question, so we simply need to find the appropriate information in the passage. From our Outline, we know that the beginning of paragraph 4 is where the criticisms of Braudel are brought up. In fact, the Label gives us everything we need: Braudel's critics believed his *scope* was *too broad*, and that the *differences* among *social sciences* were *blurred*. This prediction matches **(B)**. **(A)** is a Faulty Use of Detail. Braudel seeks structures like geohistory that underlie social activity in history; however, this is not in paragraph 4 and is not a criticism of Braudel. **(C)** is Out of Scope; while Braudel's *three temporalities* are discussed, the *relationship between them* is never addressed. Finally, **(D)** is an Opposite answer choice. According to paragraph 4, critics thought Braudel's approach *made it difficult to delimit the boundaries of observation*. This statement is at odds with *rigidly define[d] boundaries* in this answer choice.

6. C

This is an Apply question of the Example subtype. We must use the author's description of the *longue durée* to identify a similar example. Paragraph 2 defines the three temporalities and defines the *longue durée* as being an *historical wave of great length* on the order of *hundreds of years*. Only **(C)** has even remotely that long of a duration, making it the correct answer. **(A)** and **(D)** are short-lived events in history just as the *evenementielle* temporality is defined in the passage. **(B)** describes a cyclical event that *occur[s]…roughly every half-century*, which would fit the definition of the *conjonctures* temporality described in the passage.

7. A

Glancing at the answer choices, we can see that this is a Strengthen–Weaken (Beyond the Passage) question. The question stem introduces a quote from Braudel that emphasizes his interest in the influence of *one man's decisions* in history. The focus on one man's influence is most similar to the *evenementielle* temporality discussed in paragraph 2. We also see in this paragraph that Braudel considered the *longue durée* the *most fascinating of the three temporalities*. The quote given in the question stem would go against the information in the passage, so we can immediately cross off **(C)** and **(D)**, which claim that this quote would *strengthen the author's claim*. The logic presented here matches with **(A)**. **(B)** is a Distortion because the author claims that Braudel is most fascinated by the *longue durée* temporality, but that is not the same thing as saying Braudel is *not interested* at all in other aspects of history. In fact, Braudel gives the actions of great men their own temporality, the *evenementielle*, so Braudel must be at least somewhat interested in these actions.

Passage 2 (Questions 8–12)

Sample Passage Outline

P1. Balance of power: key is maintaining equilibrium

P2. Needs independent nations, proximity, and level playing field

P3. Factors needed in framework: ≥3 states, common culture, diplomacy

P4. Also need shared information on other states (power, economy, security)

P5. Neighbor analogy

Goal: To describe the conditions needed for balance of power

8. D

In this Detail question, we need to pick the choice that matches the explanation for this claim in the passage. The author clarifies the phrase *common ground of culture* by pointing out that states with similar culture are *more likely to align in cooperative manner*. This prediction is almost identical to **(D)**. **(A)** is a Faulty Use of Detail; while *mechanisms of diplomacy* are needed for a balance of power according to the author, they are a separate, third, necessary factor mentioned in this paragraph (in addition to having at least three states and common culture) and are not part of the *common ground of culture* described. **(B)** is Out of Scope because the passage does not have any discussion of cultural differences causing military conflict—it just states that cultural similarities foster cooperative behavior. **(C)** is similarly Out of Scope because the author does not discuss the influence of a hegemon on the culture of surrounding states.

9. C

This Inference question wants us to interpret a phrase used by the author. The phrase is found in the last sentence of paragraph 2. Here, the author claims that states *commensurate*, or nearly equal, in power will decide policy on a *level playing field.* This sentence contrasts with the previous one, which claims that *When one state far outstrips its close neighbors in power, it dictates economic and military policy for the region.* Taking these together, we can infer that roughly equal states will all play a role in *dictat[ing] economic and military policy for the region.* This prediction matches closely with **(C)**. **(A)** and **(B)** are subtle Distortions. While the author says that states will decide policy on a level playing field, he or she does not state that this occurs *democratic[ally]* or *to the advantage of all member states.* Notice that these first three answer choices are all very similar in that they suggest that each member state gets a say in policy-making—but **(C)** is the mildest answer choice that strays the least from what is stated in the passage. Generally, an answer choice that stays closer to what is written in the passage is a safer bet on the MCAT. Finally, **(D)** is definitely Out of Scope as the passage does not address the interaction between states within the balance of power and those *outside the balance of power.*

10. B

This is a Scattered Inference question, so find the related information in the passage. Statements I and II are both brought up in the same sentence in paragraph 4: *a nation's economic planning should consider the dynamics of the entire state system in addition to its own agriculture and industry.* In other words, a nation should *plan its agricultural and industrial development*, Statement I, while also taking into consideration the other nations in the balance of power, which would necessarily require *possess[ing] information on [their] economies*, Statement II. Thus, we can eliminate **(A)**. Statement III is not supported by the passage, however. The author clearly believes military intelligence is an important aspect of the balance of power and that states should seek out military information about other member states, but the author does not say that states ought to willingly give away military secrets to other states. This idea is a Distortion. With Statement III eliminated, the answer is **(B)**.

11. B

The word *assumption* immediately lets us know this is an Inference question of the Assumption subtype. Here, we want to identify an assumption the author makes while forming the analogy in paragraph 5. As discussed in Chapter 6 of *MCAT CARS Review*, analogical reasoning relies on two pieces of evidence: that the two entities (neighbors, N, and international balances of power, I) share similar corresponding characteristics (N_1 and I_1, N_2 and I_2, and so on), and that one of the entities has an additional characteristic (N_x)—from this, the conclusion is drawn that the other entity also has that characteristic (I_x). This might sound like a bunch of jargon, but it provides a quick way to answer this question: one assumption in an analogy is *always* that the two entities are similar enough to be compared in this way. Therefore, we should look for an answer that, were it not true, would mean that neighbors and states are actually not so similar. **(B)** matches this assumption—if *power relations between neighbors* were not *dependent on similar factors to those between neighbor states*, then the comparison made in the last paragraph would make no sense. **(A)** is a Distortion; while it makes sense that both neighbors and states would aim to *avoid conflict whenever possible*, the word *must* makes this answer Extreme. Even if neighbors occasionally quarreled, and neighbor states occasionally fought, the analogy would still hold. In fact, the analogy even suggests that a conflict may occur over the lawnmower. **(C)** is also Extreme; while the author suggests that cultural similarity facilitates cooperation, there is no suggestion that this similarity is absolutely *required for peaceful coexistence*. Finally, **(D)** is Out of Scope. The author seems to suggest that force may be needed to maintain a balance of power but does not make any reference to what is *justifiable*. Note the use of *always* also makes this answer choice Extreme.

12. A

This is a *Reasoning Within the Text* question that does not neatly fit into any category as it is asking us to extend the analogy given in paragraph 5 a bit more. In this case, we will have to take the characteristics of balances of power described paragraphs 2–4 and translate them to the neighbor example. Note that we want the *LEAST* important characteristic, implying that three answer choices are things the author would deem important for a neighborhood's balance of power. The Outline provides all of the information we need: in paragraph 3, the author states that ≥*3 states* are required in the balance of power—which is less than the number given in **(A)**. This makes it the correct answer; an alliance network could have fewer than five neighbors (specifically, three or four neighbors) and still have a balance of power. The other answer choices are all stated in the passage. **(B)** is mentioned in paragraph 3, as *skilled diplomats*—a sensible correlate to *skilled negotiators*—are needed in the balance of power. **(C)** and **(D)** are actually stated directly in the passage: *for the continued survival of the balance of power…it is critical that you find out whether your neighbor possesses a gun and whether he might use it under such circumstances [of conflict].*

Question Types I: *Foundations of Comprehension* Questions

9

9: Question Types I

In This Chapter

9.1 Main Idea Questions	**196**	**9.4 Definition-in-Context**	
Sample Question Stems	196	**Questions**	**205**
Strategy	196	Sample Question Stems	206
Worked Example—		Strategy	206
A Psychology Passage	197	Worked Example—	
9.2 Detail Questions	**198**	A History Passage	207
Sample Question Stems	199	**Concept and Strategy Summary**	**210**
Strategy	199	**Worked Example**	**212**
Worked Example—			
An Arts Passage	200		
9.3 Function Questions	**203**		
Sample Question Stems	203		
Strategy	203		
Worked Example—			
An Ethics Passage	204		

Introduction

> **LEARNING GOALS**
>
> After Chapter 9, you will be able to:
>
> - Identify Main Idea, Detail, Function, and Definition-in-Context questions
> - Solve *Foundations of Comprehension* questions with strategies based on question type
> - Recognize common features of *Foundations of Comprehension* questions

In order to get into college, you likely had to take an exam like the SAT® or ACT®, both of which feature sections that test reading comprehension. Most of the questions in these sections were straightforward, requiring you merely to search the text for a key fact, to define the meaning of a term used in a passage, or to identify the author's thesis. Some of the more challenging questions may have required you to understand what the author was *doing* with a part of the passage, to imagine things from the writer's perspective, or to explain why she used a certain word or phrase. Such questions can also be found in the *Critical Analysis and Reasoning Skills* (CARS) section of the MCAT. However, these will constitute only approximately 30 percent of the questions you encounter on Test Day, and they will take more difficult forms than you saw on those precollege exams. Our research of all of

MCAT Critical Analysis and Reasoning Skills

the released MCAT material indicates that Detail questions make up about half of the *Foundations of Comprehension* questions, or about eight or nine questions. The other question types in this category show up in about two or three questions each.

This chapter and the next two will follow the same general pattern. For each question type, we will briefly discuss what makes that type distinctive before examining some sample question stems. Then, after discussing strategy, we'll revisit one of the passages first introduced in Chapter 7 of *MCAT CARS Review*, providing at least one sample question of each type. The question types in this chapter all fall under the *Foundations of Comprehension* category and will be examined in this order: Main Idea, Detail, Function, and Definition-in-Context.

Note: The Question Types, as well as the Kaplan Method for CARS Passages, Kaplan Method for CARS Questions, and Wrong Answer Pathologies are included as tear-out sheets in the back of this book.

9.1 Main Idea Questions

> **MCAT Expertise**
>
> According to our research of released AAMC material, Main Idea questions make up about 5 percent of the CARS section (about two or three questions).

Questions that ask about the big picture of the passage are what we call **Main Idea questions**. Only about 5 percent of the questions in the CARS section fall into this type, but they are easy to recognize and the Kaplan Method for CARS Passages arms you with everything you need to attack them.

SAMPLE QUESTION STEMS

- The author's central thesis is that:
- Which of the following best characterizes the main idea of the passage?
- The primary purpose of the passage is to:
- The author of the passage is primarily interested in:
- Which of the following titles best captures the main theme of the passage?
- The author can best be viewed as a proponent of:
- The language used in the passage makes it clear that the intended audience is:
- What is the author's central concern?

> **Bridge**
>
> Main Idea questions that ask about the audience or the medium are checking your rhetorical analysis skills. Analysis of the likely author, her tone, and her voice can reveal the intended audience and most likely medium. Rhetoric was discussed in Chapter 2 of *MCAT CARS Review*.

Main Idea questions will either use some variant of the phrase *central thesis*, *primary purpose*, or—of course—*main idea*, or make some kind of general reference to *the author*. On rare occasions, Main Idea questions will bring in a more challenging aspect of the rhetorical situation, such as the *audience* or the *medium*.

STRATEGY

Decide to do a Main Idea question as soon as you encounter it because these questions can gain you some quick points. Your best bet for a Plan is simply to *Go for the Goal!* In fact, this is one of the reasons it's worth Reflecting on the passage at the end of your Reading: the Goal that you write down in your Outline gives you an instant prediction for questions of this type. Note that the verb you choose will often be as important a part of your prediction as any of the words that follow.

In the event that none of the choices come close to matching what you thought the author's Goal was, use process of elimination to remove Faulty Uses of Detail that are too narrow, Out of Scope options that go too far afield, and any choice that has the wrong tone (positive, negative, ambivalent, or impartial) or degree (extreme *vs.* moderate).

WORKED EXAMPLE—A PSYCHOLOGY PASSAGE

There is no shortage of evidence for the existence of systemic biases in ordinary human reasoning. For instance, Kahneman and Tversky in their groundbreaking 1974 work proposed the existence of a heuristic—an error-prone shortcut in reasoning—known as "anchoring." In one of their most notable experiments, participants were exposed to the spin of a roulette wheel (specially rigged to land randomly on one of only two possible results) before being asked to guess what percentage of United Nations member states were African. The half of the sample who had the roulette wheel stop at 65 guessed, on average, that 45% of the UN was African, while those with a result of 10 guessed only 25%, demonstrating that prior presentation of a random number otherwise unconnected to a quantitative judgment can still influence that judgment.

The anchoring effect has been observed on repeated other occasions, such as in Dan Ariely's experiment that used digits in Social Security numbers as an anchor for bids at an auction, and in the 1996 study by Wilson *et al.* that showed even awareness of the existence of anchoring bias is insufficient to mitigate its effects. The advertising industry has long been aware of this bias, the rationale for its frequent practice of featuring an "original" price before showing a "sale" price that is invariably reduced. Of course, anchoring is hardly alone among the defective tendencies in human reasoning; other systemic biases have also been experimentally identified, including loss aversion, the availability heuristic, and optimism bias.

Example:

1. The author's primary task in the passage is to:
    A. search for evidence of systemic biases in normal human thinking.
    B. discuss empirical findings on anchoring and other reasoning biases.
    C. show that anchoring is the most commonly occurring error in reasoning.
    D. demonstrate that knowledge of anchoring bias is insufficient to prevent it.

MCAT Expertise

The wrong answer choices in Main Idea questions are very predictable. One or more tend to be too narrow, reflecting the ideas from only one paragraph. One or more tend to be too broad, becoming Out of Scope. One or more tend to embody the wrong tone (positive, negative, ambivalent, or impartial) or degree (extreme *vs.* moderate).

MCAT Critical Analysis and Reasoning Skills

> **Solution:** The phrase *primary task* tells you that this is a Main Idea question, so let's look again at the Outline we generated for this brief passage in Chapter 7 of *MCAT CARS Review*:
>
> **P1.** Systemic biases in reasoning: ex. anchoring heuristic (evidence: Kahneman & Tversky)
>
> **P2.** More evidence for anchoring (Ariely, Wilson *et al.*); other systemic biases
>
> **Goal:** To present evidence for systemic biases in reasoning, especially anchoring
>
> Using that Goal, or the Goal you generated during your own Outline attempt, your prediction should immediately lead you to **(B)**. The phrase *discuss empirical findings* is roughly equivalent in meaning to our *present evidence*. Also, the scope is right, with the answer focusing on anchoring first but mentioning other biases as well because the author does introduce anchoring as an example with the phrase *For instance*.
>
> On Test Day, whenever you find an easy match for a prediction that you made or a choice that fulfills your expectation for correctness, go with that Answer and move on to Assess the next question. But for these examples, it's worth discussing what goes wrong in the other options. **(A)** comes close and has the appropriate scope, but the verb is wrong: in the very first sentence, the author says that *There is no shortage of evidence*, so why would the author's task be to *search for* something so readily available? Moving on to **(C)**, we find an Out of Scope choice: while the author says that systemic biases are common to thinking, there is never any comparison among the types mentioned. Just because the author focuses on anchoring does *not* mean that anchoring is necessarily the most frequently occurring—the author could choose to discuss it for any number of reasons other than its commonness. Finally, **(D)** is a Faulty Use of Detail because it is too specific. The referenced claim is made in the second paragraph, but it applies only to the study conducted by Wilson and others. This answer does not encompass the focus of the whole passage.

9.2 Detail Questions

Detail questions ask about what is stated explicitly in the passage. These are probably what you typically think of when you imagine a "reading comprehension" question because they tend to require searching the text to fill in the missing piece. While Main Idea questions focus on the big picture, Detail questions zoom in on some of the finer points of the passage. They are by far the most common type of the *Foundations of Comprehension* category, constituting at least half of the questions that fall under this heading.

MCAT Expertise

According to our research of released AAMC material, Detail questions make up about 16 percent of the CARS section (about eight or nine questions).

SAMPLE QUESTION STEMS

- According to the author's account of [topic], [concept] is:
- The author states that [person] holds the view that:
- Which of the following, according to the passage, does the author associate with the idea of [concept]?
- The author's apparent attitude toward [alternative position] is:
- Based on the discussion in [paragraph reference], the work of [artist/writer] was widely regarded as:
- The passage suggests which of the following about [topic]? [list of Roman numerals]
- The author asserts all of the following EXCEPT:
- Which of the following claims does NOT appear in the passage?

Detail questions tend to contain simple, declarative language (*is* and *are*) rather than the subjunctive mood (*would* and *could*), often include phrases like *the author states* and *according to the passage*, and more often than not take the form of incomplete sentences ending with a colon.

The last three examples listed fall into what is known as the Scattered subtype. A **Scattered** question is one that either employs a set of Roman numeral options or uses a word like EXCEPT, NOT, or LEAST. While just about any type of question can be Scattered—from Scattered Function to Scattered Inference—Scattered Detail is perhaps the most common example of the Scattered subtype.

STRATEGY

The only trick to working with Detail questions is that sometimes a seemingly straightforward question can actually require making an inference. If you are dealing with a true Detail question, though, your Plan should be to follow the clues in the question stem, especially the content **buzzwords**, looking to your Outline to help you home in on the relevant portion of the passage. Once you find the precise sentence referenced, remember to read at least the sentence before and the sentence after (unless of course you're looking at the first or last sentence of the entire passage!) to get a bit more context. Once you've read the relevant text, create your prediction by putting the sentence in your own words, and then look for the best match to Answer the question.

You will likely want to save the Scattered Detail questions you encounter for later because these will often require researching three or four different parts of the text, rather than just one. When working with these questions, your Outline will prove extremely useful for locating everything you need to find. Process of elimination is almost inevitable with questions of the Scattered subtype, but that doesn't mean you

Bridge

Sometimes, what appears to be a Detail question will actually require you to make an inference. Inference questions include both Assumption and Implication questions, the strategies for which are discussed in Chapter 10 of *MCAT CARS Review*.

MCAT Expertise

For Detail questions, make sure to paraphrase the relevant text in a "short and sweet" format that will be easy to repeat to yourself while reading the answer choices. Much of the challenge of this type of question can be trying to figure out which answer choice actually matches your prediction—so make your prediction something that's easy to remember!

MCAT Critical Analysis and Reasoning Skills

should immediately jump to the answer choices. As with any question, take what you can from the stem to set some basic expectations. You have your Outline, as well as the experience you will have gained working on other questions in the set, to help with this.

WORKED EXAMPLE—AN ARTS PASSAGE

One of the first examples of the ascendance of abstraction in 20th-century art is the Dada movement, which Lowenthal dubbed "the groundwork to abstract art and sound poetry, a starting point for performance art, a prelude to postmodernism, an influence on pop art…and the movement that laid the foundation for surrealism." Dadaism was ultimately premised on a philosophical rejection of the dominant culture, which is to say the dominating culture of colonialist Europe. Not content with the violent exploitation of other peoples, Europe's ruling factions once again turned inward, reigniting provincial disputes into the conflagration that came to be known by the Eurocentric epithet "World War I"—the European subcontinent apparently being the only part of the world that mattered.

The absurd destructiveness of the Great War was a natural prelude to the creative absurdity of Dada. Is it any wonder that the rejection of reason made manifest by senseless atrocities should lead to the embrace of irrationality and disorder among the West's subaltern artistic communities? Marcel Janco, one of the first Dadaists, cited this rationale: "We had lost confidence in our culture. Everything had to be demolished. We would begin again after the *tabula rasa*." Thus, we find the overturning of what was once considered art: a urinal becomes the *Fountain* after Marcel Duchamp signs it "R. Mutt" in 1917, the nonsense syllables of Hugo Ball and Kurt Schwitters transform into "sound poems," and dancers in cardboard cubist costumes accompanied by foghorns and typewriters metamorphosize into the ballet *Parade*. Unsurprisingly, many commentators, including founding members, have described Dada as an "anti-art" movement. Notwithstanding such a designation, Dadaism has left a lasting imprint on modern Western art.

> ### Example:
>
> 2. As stated in the passage, prior to the Great War, the leaders of Europe were primarily focused on:
>   A. fighting one another in World War I.
>   B. colonizing other parts of the globe.
>   C. gazing inward at local problems.
>   D. rejecting the dominant culture.

Solution: When the stem says *as stated in the passage*, it usually signifies a Detail question (with some exceptions, noted in the next chapter). You'll most likely want to work on it right away. There are two buzzwords in the question stem: *the Great War* and *the leaders of Europe*. The first comes directly from the start of paragraph 2, where you can judge from the context that it must be another name for what we now call World War I. (They of course didn't know there was going to be a second one at the time!) The second buzzword does not appear verbatim, but it shows up in the synonymous phrase *Europe's ruling factions* in the last sentence of paragraph 1. In addition to reading these two sentences, it may be worth reading one before and one after.

In fact, the preceding sentence offers a key bit of context, illustrating that the author regards European culture quite negatively, with the phrase *the dominating culture of colonialist Europe*. More of the same follows with the phrase *violent exploitation of other peoples*, as this author highlights Europe's past as a global colonizer. The phrase *once again turned inward* is noteworthy for mirroring the language of *focused on* featured in the question stem. Putting this all together sets a thorough expectation for the correct answer: if the turning inward happened with the Great War, then prior to that, Europe's leaders must have been looking outward, focusing on their colonial acquisitions. This prediction finds a perfect match in **(B)**.

Among the wrong options, **(A)** is based on a misunderstanding of the terminology. Though World War I is mentioned in the prior sentence, the author calls attention to the fact that it only later came to be known by that *epithet* (name) and, then in the following sentence, uses another name for it. Because the two terms refer to the same event, European leaders could not be focused on it before it happened. In contrast, **(C)** is an Opposite because Europeans were actually turning outward at their colonies. Finally, **(D)** is a Faulty Use of Detail: that phrase appears almost exactly, but it actually describes the founders of the Dada movement, not the European leaders.

Example:

3. Based on the passage, which of the following is not a characteristic associated with Dadaism?
    A. A renunciation of European culture
    B. A reputation as an "anti-art" movement
    C. Importance for later 20th-century art
    D. The embrace of irrational atrocities

Solution: Although the testmaker will often put the word *not* in italics or all caps, sometimes it will appear plainly in a question stem, as seen here. Don't be fooled: this is still a Scattered Detail question. Because Scattered Detail questions are often time-consuming, you'd likely want to make it the last question you work on in the question set. That said, don't forget to make a Plan and set some basic expectations. The phrasing *a characteristic associated with* is vague enough that it could mean something the author says or potentially another view identified in the passage. That means you can't rule out an answer choice just because it doesn't sound like something the author would say, so long as it sounds like something someone else in the passage would. Because both paragraphs are chock-full of claims about Dadaism, there's no point in searching the text to set additional expectations—you'll have to go with this more minimal Plan.

(A) comes directly from the author's second sentence: *Dadaism was ultimately premised on a philosophical rejection of the...culture of colonialist Europe*. The word *renunciation* is a synonym of *rejection*—so cross off this option. Moving on to the next choice, while the author explicitly rejected the idea that Dadaism is "anti-art," that view is still reflected in the passage in the next-to-last sentence, where the author writes *many commentators, including founding members, have described Dada as an "anti-art" movement*. Another way of saying that many commentators describe it that way is to say that it has a reputation. Hence, **(B)** should also be eliminated. The next characteristic is found at both the beginning of the passage, with the quotation from Lowenthal, and at the end, with the author's statement that *Dadaism has left a lasting imprint on modern Western art*. With **(C)** now off the table, we know that **(D)** must be the answer—but let's check it anyway.

This choice is very close to being a characteristic; however, the word *atrocities* prevents it from being associated with Dadaism and makes it the correct choice. The key sentence is worth repeating in full in order to untangle it: *Is it any wonder that the rejection of reason made manifest by senseless atrocities should lead to the embrace of irrationality and disorder among the West's subaltern artistic communities?* The author is saying that Dadaism embraces irrationality, although not in the same way that irrationality (*rejection of reason*) manifests itself in warfare (*senseless atrocities*). The preceding sentence offers some useful clarification: *The absurd destructiveness of the Great War was a natural prelude to the creative absurdity of Dada*. Notice how the word *absurd* is used to join two concepts that are typically viewed as opposites: creation and destruction. The author is suggesting that, even though both the Great War and Dadaism defied reason, they did so in dramatically different ways. This irrationality or absurdness in Dadaism stood in opposition to the atrocities of World War I; the word *embrace* in the answer choice makes this statement *not* a characteristic of Dadaism. That's why **(D)** is correct.

9.3 Function Questions

One of the reasons that the Kaplan Method for CARS Passages emphasizes reading for perspective (trying to understand the author's attitude and intentions) is that the entire **Function question** type specifically asks about what the author is trying to *do* in the passage. Unlike a Main Idea question, which might ask about the overall Goal of the passage, a Function question will ask about the purpose of only a portion of it. Function questions are just about as common as Main Idea questions, constituting about 5 percent of what you'll see in this section on Test Day.

SAMPLE QUESTION STEMS

- What is the author's apparent purpose in stating [quotation]?
- The author mentions [topic] in [paragraph reference] in order to:
- Which of the following is the most probable reason for the author's inclusion of a quotation from [person]?
- The author's reference to [concept] in [paragraph reference] is most likely supposed to show:
- When the author says [claim], she is emphasizing that:
- Which of the following is the example from [paragraph reference] most likely intended to suggest?
- The author compares [one concept] to [another concept] because:
- The author's principal motive for discussing [alternative position] is to explain that:

What should be readily apparent in the phrasing of Function questions is frequent mention of the author and the use of direct references to the text—especially through paragraph references. Language like *purpose*, *motive*, and *intention* indicate a Function question, as do phrases that end with *in order to* and *because*.

STRATEGY

The Outline will be key when working with a Function question. If you were reading for perspective by looking out for Author keywords (which give a glimpse from the author's point of view), you may have already included the information you need to answer such a question in your Outline. Keep in mind that, generally speaking, Function questions work in a nested way. In other words, the passage as a whole has a purpose, and each paragraph within it has a subordinate function that is distinctive but that contributes to the larger whole. Each paragraph can in turn be broken down into sentences, each of which has its own particular role to play in the paragraph—and even sentences can be broken down into particular words or phrases.

> **MCAT Expertise**
>
> According to our research of released AAMC material, Function questions make up about 5 percent of the CARS section (about two or three questions).

MCAT Critical Analysis and Reasoning Skills

> **Key Concept**
>
> Both Main Idea and Function questions can often be answered solely by looking at your Passage Outline. The answers to Main Idea questions should reflect the author's Goal, whereas the answers to Function questions should usually reflect the Label you've assigned to a given paragraph.

Because a Function question will generally ask for the purpose of no more than a paragraph, to formulate some initial expectations, you should go to your Outline and look at the Label for the specific paragraph and perhaps also look at the Goal. Then, if buzzwords in the question stem direct you to specific sentences, reread those portions of the paragraph and think about how they fit into both the purpose of the paragraph and the passage's general Goal. Formulate a statement of what the desired function is, and then start to look for an Answer that matches.

Remember that if you can't find a perfect match, you can eliminate choices that would be inconsistent with functions at a higher level. So, for example, the purpose of a paragraph will not be at odds with the author's Goal for the passage as a whole. Even when authors bring up information that conflicts with their main arguments, they commonly do so for the sake of shooting it down—answering or countering a refutation, as discussed in Chapter 5 of *MCAT CARS Review*.

WORKED EXAMPLE—AN ETHICS PASSAGE

The most prevalent argument against doctor-assisted suicide relies upon a distinction between *passive* and *active* euthanasia—in essence, the difference between killing someone and letting that person die. On this account, a physician is restricted by her Hippocratic oath to do no harm and thus cannot act in ways that would inflict the ultimate harm, death. In contrast, failing to resuscitate an individual who is dying is permitted because this would be only an instance of refraining from help and not a willful cause of harm. The common objection to this distinction, that it is vague and therefore difficult to apply, does not carry much weight. After all, applying ethical principles of *any sort* to the complexities of the world is an enterprise fraught with imprecision.

Rather, the fundamental problem with the distinction is that it is not an ethically relevant one, readily apparent in the following thought experiment. Imagine a terminally ill patient hooked up to an unusual sort of life support device, one that only functioned to prevent a separate "suicide machine" from administering a lethal injection so long as the doctor pressed a button on it once per day. Would there be any relevant difference between using the suicide machine directly and not using the prevention device? The intention of the doctor would be the same (fulfilling the patient's wish to die), and the effect would be the same (an injection causing the patient's death). The only variance here is the means by which the effect comes about, and this is not an ethical difference but merely a technical one.

9: Question Types I

> **Example:**
>
> 4. The author's apparent intention in discussing the "suicide machine" in paragraph 2 is to:
>   A. support his thesis using an imaginative exercise.
>   B. question the idea that vagueness is ethically relevant.
>   C. explain the operation of a piece of medical equipment.
>   D. propose a new method for performing euthanasia.
>
> **Solution:** As with the Main Idea example above, looking at our Outline for this passage from Chapter 7 of *MCAT CARS Review* will make this question much more manageable:
>
> **P1.** "Most prevalent argument" against euthanasia assumes passive/active difference (vague is okay)
>
> **P2.** Author: distinction is not ethically relevant (supported w/thought experiment)
>
> **Goal:** To argue the distinction between passive and active euthanasia is not ethically relevant
>
> The purpose listed for paragraph 2 even includes a note about the *"suicide machine"* thought experiment's role as support, so this will suffice as a prediction. Going back to the text shouldn't be necessary unless the answer choices take us somewhere that we completely didn't expect. Fortunately, this is not the case with this question, and we can see that **(A)** gives us precisely what we need. That time spent Outlining has been more than paid back with a quick correct response.
>
> Among the wrong answers, **(B)** is wrong for bringing in *vagueness*, when the point of the thought experiment is to question whether the *passive/active distinction* is ethically relevant—not vagueness. **(C)** might be considered a Faulty Use of Detail because the operation of this machine is explained. However, the machine is being described not for its own sake—it's imaginary, after all—but simply to illustrate a point. The final incorrect option, **(D)**, would be a product of taking the thought experiment too literally.

9.4 Definition-in-Context Questions

Definition-in-Context questions constitute the final type that falls into the *Foundations of Comprehension* category. The task involved with such questions is always the same: define the word or phrase, specifically as it is used in the passage. You can expect to see about two of them on Test Day, as they tend to make up about 4 percent of all questions in the CARS section.

> **MCAT Expertise**
>
> According to our research of released AAMC material, Definition-in-Context questions make up about 4 percent of the CARS section (about two questions).

205

MCAT Critical Analysis and Reasoning Skills

SAMPLE QUESTION STEMS

- As used by the author, the word [term] most nearly means:
- In [paragraph reference], what is the author's most likely meaning when stating [quotation]?
- The author's choice of the phrase [term] is probably intended to suggest:
- As used in the passage, [term] refers to:
- In [paragraph reference], the author asserts that [claim]. What the author most likely means by this is:
- Which of the following is most synonymous with [concept] as discussed in the text?

This list of stems makes it clear that Definition-in-Context questions always feature a reference to a word, phrase, or an entire claim from the passage, the meaning of which you are tasked with identifying. Quotation marks and italics are common features used to call attention to the terms, but on occasion a Definition-in-Context question stem may lack these.

STRATEGY

Although these questions ask about the meanings of words, a dictionary will not help you here, and in some cases it could even lead you astray. Trap answers in these questions are often the common definitions of the word, which are tempting Out of Scope choices that fail to match the context. That said, these questions tend to be relatively quick because they refer only to small portions of the text. Thus, you should generally decide to work on these questions as soon as you see them.

Your Plan with a Definition-in-Context question will be to go to the text and surrounding context, if necessary, to see how the word or phrase is actually used in the passage. With this question type, the question stems will usually contain a paragraph reference, but use your Outline if necessary to locate the relevant sentence. If reading that sentence doesn't give you enough to work with, then look at the one before and after. Then phrase a definition in your own words based on what you see. Author keywords may be especially helpful because answer choices with the wrong tone can immediately be ruled out.

MCAT Expertise

An author may imbue common words with a special meaning in the passage. Therefore, make sure to check how the author actually uses the word in a Definition-in-Context question, rather than looking for a dictionary definition of the term. Wrong answers in these questions are often accurate definitions of the term that do not match how the term was used in the passage.

9: Question Types I

WORKED EXAMPLE—A HISTORY PASSAGE

In 1941, an exuberant nationalist wrote: "We must accept wholeheartedly our duty and our opportunity as the most powerful and vital nation...to exert upon the world the full impact of our influence, for such purposes as we see fit and by such means as we see fit." If forced to guess the identity of the writer, many US citizens would likely suspect a German jingoist advocating for *Lebensraum*. In actuality, the sentiment was expressed by one of America's own: Henry Luce, the highly influential publisher of the magazines *Life*, *Time*, and *Fortune*. Luce sought to dub the 1900s the "American Century," calling upon the nation to pursue global hegemony as it slipped from the grasp of warring Old World empires. As a forecast of world history, Luce's pronouncement seems prescient—but is it justifiable as a normative stance?

Not all of Luce's contemporaries bought into his exceptionalist creed. Only a year later, Henry Wallace, vice president under FDR, insisted that no country had the "right to exploit other nations" and that "military [and] economic imperialism" were invariably immoral. It is a foundational assumption in ethics that the wrongness of an act is independent of the particular identity of the actor—individuals who pay no heed to moral consistency are justly condemned as hypocrites. So why should it be any different for nation-states? In accord with this principle, Wallace proselytized for "the century of the common man," for the furtherance of a "great revolution of the people," and for bringing justice and prosperity to all persons irrespective of accidents of birth. Sadly, Wallace never had the chance to lead the United States in this cosmopolitan direction; prior to Roosevelt's demise at the beginning of his fourth term, the vice presidency was handed to Harry Truman, a man whose narrow provincialism ensconced him firmly in Luce's camp. And with Truman came the ghastly atomic eradication of two Japanese cities, the dangerous precedent set by military action without congressional approval in Korea, and a Cold War with the Soviet Union that brought the world to the brink of nuclear destruction.

Example:

5. The author's use of the term "provincialism" in paragraph 2 comes closest in meaning to:
   A. German jingoism.
   B. economic imperialism.
   C. nationalistic exceptionalism.
   D. exuberant cosmopolitanism.

MCAT Critical Analysis and Reasoning Skills

> **Solution:** The clue in the question stem points to the second paragraph, but the reference might still be difficult to find in a paragraph full of *–ism*s. The key sentence is this: *Sadly, Wallace never had the chance to lead the United States in this cosmopolitan direction; prior to Roosevelt's demise at the beginning of his fourth term, the vice presidency was handed to Harry Truman, a man whose narrow provincialism ensconced him firmly in Luce's camp.* This sentence gives you a lot to work with. You can see that *provincialism* is a view attributed to Truman and Luce, who are contraposed to the *cosmopolitan* Wallace. In a passage that features a few different perspectives, it's not surprising to see a question like this that requires keeping straight who holds which view. This passage presents two major sides, and it's clear that *provincialism* represents a view that belongs to Truman, whose views align with those of Luce.
>
> With these expectations established, look for a match in the answers. Because both nationalism (at the start of paragraph 1) and exceptionalism (at the start of paragraph 2) are views attributed to Luce, it is evident that **(C)** must be correct. **(A)** is a Faulty Use of Detail; *a German jingoist* was mentioned in the first paragraph, but this phrase is no good because Luce and Truman are both Americans. **(B)** echoes the quotation from Wallace as he criticizes Luce—*"military [and] economic imperialism"*—but all of the examples given for Truman show evidence of military action, not economic: *the ghastly atomic eradication of two Japanese cities, the dangerous precedent set by military action without congressional approval in Korea, and a Cold War with the Soviet Union that brought the world to the brink of nuclear destruction*. The emphasis in **(B)** is therefore misplaced. Lastly, **(D)** uses a variant of *cosmopolitan*, a characteristic that was attributed to Wallace, so it's an Opposite.

Conclusion

Reading comprehension is a skill you've honed through your whole life—most exams you've had in literature and English classes have most likely centered on your ability to understand the text you read. On the MCAT, reading comprehension is important not only for answering *Foundations of Comprehension* questions but also for understanding the passage itself. While they take many forms, including Main Idea, Detail, Function, and Definition-in-Context questions, all *Foundations of Comprehension* questions share a few common features. Answering all of these question types is facilitated by a well-constructed Outline, which can be used to locate the relevant text for the answer or—in the case of Main Idea and some Function questions—will contain the answer itself. Also, the answers to all of these questions must be stated or paraphrased directly in the passage. These question types also differ in significant ways. For example, Main Idea and Definition-in-Context questions should usually be answered as soon as you see them, whereas Detail (especially Scattered Detail) questions may be more time-consuming and should therefore be saved until the end of the question set. Each question type has a specific Plan, as described in this chapter. We'll continue discussing question types in the next two chapters as we explore *Reasoning Within the Text* and *Reasoning Beyond the Text* questions.

MCAT Critical Analysis and Reasoning Skills

CONCEPT AND STRATEGY SUMMARY

Main Idea Questions

- Assess: **Main Idea questions** ask for the author's primary Goal.
    - These questions often contain words like *central thesis*, *primary purpose*, or *main idea*.
    - Less commonly, these questions may ask about a different aspect of the rhetorical situation such as the *audience* or the *medium*.
- Plan: Look at what you wrote in your Outline for the Goal.
- Execute: Reread the Goal in your Outline, taking note of the charge and degree of the verb (positive *vs.* negative, extreme *vs.* moderate).
- Answer: Match your expectations with the right answer. If there is no clear match, or if you cannot perform any of the earlier steps of the Kaplan Method for CARS Questions, use process of elimination.
    - Wrong answer choices may be too narrow (Faulty Use of Detail) or too broad (Out of Scope).
    - Wrong answer choices may have the wrong tone (positive, negative, ambivalent, or impartial) or degree (too extreme or too moderate).

Detail Questions

- Assess: **Detail questions** ask about what is stated explicitly in the passage.
    - These questions tend to use words like *the author states* or *according to the passage*, with declarative language like *is* and *are*.
    - Detail questions are the most likely to use the **Scattered** format, which uses Roman numeral options or words like *EXCEPT*, *NOT*, or *LEAST*.
- Plan: Look for content buzzwords in the question stem, and check your Outline to determine where the relevant information will be found.
- Execute: Reread the relevant sentence as well as the sentences before and after. Create your prediction by putting the answer in your own words.
    - Make the prediction brief so you can repeat it to yourself between answer choices.
    - For Scattered Detail questions, locate all three of the wrong answers in the passage so you can eliminate them from the options.
- Answer: Match your expectations with the right answer. If there is no clear match, or if you cannot perform any of the earlier steps of the Kaplan Method for CARS Questions, use process of elimination.

210

Function Questions
- Assess: **Function questions** ask about what the author is trying to *do* during the passage.
  - These questions are similar to Main Idea questions, although they focus on the purpose of only one portion of the passage (usually one sentence or one paragraph).
  - Function questions tend to use words like *purpose*, *motive*, or *intention* or phrases like *in order to* or *because*.
- Plan: Use your Outline to locate the relevant paragraph.
- Execute: Look at your Label for the relevant paragraph and at the Goal at the bottom of your Outline. If buzzwords in the question stem direct you to specific sentences, reread those portions, thinking about how they fit into the purpose of the paragraph and the overall passage.
- Answer: Match your expectations with the right answer. If there is no clear match, or if you cannot perform any of the earlier steps of the Kaplan Method for CARS Questions, use process of elimination, removing any answer that conflicts with the author's main argument or the paragraph's purpose.

Definition-in-Context Questions
- Assess: **Definition-in-Context questions** ask you to define a word or phrase as it is used in the passage.
  - These questions often call attention to the term to be defined using quotation marks or italics, but not always.
  - Definition-in-Context questions always reference a word, phrase, or an entire claim from the passage.
- Plan: Use your Outline to locate the relevant paragraph.
- Execute: Reread the sentence with the word or phrase, and perhaps the surrounding context. Rephrase the author's definition of the term in your own words.
- Answer: Match your expectations with the right answer. If there is no clear match, or if you cannot perform any of the earlier steps of the Kaplan Method for CARS Questions, use process of elimination.

MCAT Critical Analysis and Reasoning Skills

WORKED EXAMPLE

Use the Worked Example below, in tandem with the subsequent practice passages, to internalize and apply the strategies described in this chapter. The Worked Example matches the specifications and style of a typical MCAT *Critical Analysis and Reasoning Skills* (CARS) passage.

Passage	Analysis
…Until last year, many people—but not most economists—thought that economic data told a simple tale. On one side, productivity—the average output of an average worker—was rising. And although the rate of productivity increase was very slow during the 1970s and early 1980s, the official numbers said that it had accelerated significantly in the 1990s. By 1994, an average worker was producing about 20 percent more than 1978.	A Scan of the passage tells us that we're dealing with a economics passage. These tend to be long, contain dense paragraphs, and are full of technical jargon. We can take comfort here because we know that they're very focused on arguments, and much of the jargon is merely evidence supporting said arguments. If we can figure out the argument elements and tone, we should be good to go. We see *Until last year*, something about *economic data*, and the author hints that it must be more complicated than it seems. We can use this implication to help anticipate where this passage might be going. So far, we know that increasing worker productivity is at the heart of some disagreement involving economists. A Label for this paragraph could be: **P1.** Productivity increased 70s to 80s, complex economic reason?
On the other hand, other statistics said that real, inflation-adjusted wages had not been rising at the same rate. Some commonly cited numbers showed real wages actually falling over the last 25 years. Those who did their homework knew that the gloomiest numbers overstated the case…Still, even the most optimistic measure, the total hourly compensation of the average worker, rose only 3 percent between 1978 and 1994…	*On the other hand* immediately clues us in that we're about to see the other side of the argument: *wages*, at least to some degree, did not keep up. Barely two paragraphs into the passage, we have the fundamentals of the whole argument, productivity *vs.* wages. *Those who did their homework* clues us into the author's opinion—that the numbers, while bad, were not as bad as some say. We can start to anticipate where the passage as a whole may be going; the author will likely analyze how productivity and wages actually changed during the time period from 1978 to 1994. Expect that the author's argument will be ripe for questions. This paragraph can be Labeled as: **P2.** Wages barely increased, Auth: pessimistic overstatement

9: Question Types I

Passage	Analysis
…But now, experts tell us it may have been a figment of our statistical imaginations…a blue-ribbon panel of economists headed by Michael Boskin of Stanford declared that the Consumer Price Index [C.P.I.] had been systematically overstating inflation, probably by more than 1 percent per year for the last two decades, mainly failing to take account of changes in consumption patterns and product quality improvements…	We love seeing Difference keywords lead off paragraphs. We're about to counter some part of the argument. We just have to be careful that we know which argument the author is contradicting and what *it* in the first sentence refers to. In this case, experts tell us that the gloom indicated by depressed wages in paragraph 2 is likely not as bad as was thought. *Michael Boskin*'s work on the *C.P.I.* and *inflation* is also seemingly supported by the author, who refers to the group as a *blue-ribbon panel*. Language this positively descriptive, along with what we saw in the last paragraph, gives us a really good idea that the author thinks the wage data is not as bad as some people make it out to be. An appropriate Label is: **P3.** Counterargument: Boskin cites inflation/C.P.I. errors
…The Boskin Report, in particular, is not an official document—it will be quite a while before the Government actually issues a revised C.P.I., and the eventual revision may be smaller than Boskin proposed. Still, the general outline of the resolution is pretty clear. When revisions are taken into account, productivity growth will probably look somewhat higher than before because some of the revisions will also affect how we calculate growth. But the rate of growth of real wages will look much higher—roughly in line with productivity. In other words, the whole story about workers not sharing in productivity gains will turn out to have been based on a statistical illusion…	The Boskin report *is not an official document, it will be quite a while before* revision happens, and the *revision may be smaller* all give us a sense of tone. We already saw that the author tends to agree with Boskin. Here, we get a sense of the limits of applicability of Boskin's work. All of the aforementioned points suggest that some aspect of Boskin's work won't be a huge deal. Further, because productivity growth will *probably* look *somewhat* higher, the Moderating keywords tell us that Boskin's work doesn't drastically alter the productivity growth-based view from paragraph 1. However, the *growth of real wages will look much higher*, so Boskin's work will more dramatically affect the argument about wages seen in paragraph 2. Handily, *in other words* summarizes this all for us: wages will parallel productivity. We can use this summary provided by the author to verify the Goal we identified at the start of the paragraph. This paragraph could be Labeled: **P4.** With Boskin's revisions, productivity will parallel real wages

MCAT Critical Analysis and Reasoning Skills

Passage	Analysis
It is important not to go overboard on this point. There are real problems in America, and our previous concerns were not pure hypochondriasis. For one, economic progress over the past 25 years has been much slower than in the previous 25. Even if Boskin's numbers are right, median family income—which officially has experienced virtually no gain since 1973—has risen by only about 35 percent over the past 25 years, compared with 100 percent over the previous 25. Furthermore, it is likely that if we "Boskinized" the old data—that is, if we tried to adjust the C.P.I. for the 50s and 60s to take account of changing consumption patterns and rising product quality—we would find that official numbers understated the rate of progress just as much if not more than they did in recent decades…	The author explicitly tells us *not to go overboard* on the point of the previous paragraph and follows that up with the key phrase *even if Boskin's numbers are right*. We know Boskin has been used as evidence for some facets of the author's argument. At the same time, though, this paragraph lets us know that there is still some problem that can't be solved when then data is *"Boskinized."* Applying Boskin's method to even older data might justify the same sort of problem that paragraph 1 brought up. A Label for this paragraph is: **P5.** Boskin limits: income gains may still look small historically
…Moreover, while workers as a group have shared fully in national productivity gains, they have not done so equally. The overwhelming evidence of a huge increase in income inequality in America has nothing to do with price indices and is therefore unaffected by recent statistical revelations. Families in the bottom fifth, who had 5.4 percent of total income in 1970, had only 4.2 percent in 1994; over the same period, the top 5 percent went from 15.6 to 20.1. Corporate CEOs, who used to make about 35 times as much as their employees, now make 120 times as much or more…	*Moreover* tells us that the author is continuing support for the prior paragraph. Not only were income gains small, but workers have *not [shared] equally* in productivity gains and there has been a *huge increase in income inequality*. The author then includes statistics to demonstrate this point. A Label for this paragraph is: **P6.** Income inequality increased
…While these are real and serious problems, however, one thing is now clear: the truth about what is happening in America is more subtle than the simplistic morality play about greedy capitalists and oppressed workers that so many would-be sophisticates accepted only a few months ago.	*One thing is now clear* tells us that whatever comes next is important. The wage disparities laid out above don't have a *simplistic* explanation. If we had picked up on the notion of complexity in the first paragraph, this should reassure us that we predicted correctly. Had we missed the notion, this would give us a second chance to pick up on this important point. This paragraph is Labeled as: **P7.** Auth: This is a complicated situation

9: Question Types I

Here's a sample Outline and Goal for this passage:

P1. Productivity increased 70s to 80s, complex economic reason?

P2. Wages barely increased, Auth: pessimistic overstatement

P3. Counterargument: Boskin cites inflation/C.P.I. errors

P4. With Boskin's revisions, productivity will parallel real wages

P5. Boskin limits: income gains may still look small historically

P6. Income inequality increased

P7. Auth: This is a complicated situation

Goal: To argue that the C.P.I., with or without Boskin's work, understated real wages; and that the productivity/wage disparity is more complicated than it first appears

Question	Analysis
1. According to the passage, "Boskinization" adjusts the C.P.I. by:	The words *according to the passage* tell us this is a Detail question, which should be quick points. The term *"Boskinized"* appears in paragraph 5, where we find that it means that *Boskin adjusted the C.P.I. to take account of changing consumption patterns and rising product quality*. This is a solid prediction.
A. increasing wages and decreasing productivity to reconcile the present disparity.	**(A)** may be tempting because Boskin's model did, in the end, increase apparent wages, but the passage makes no mention of *decreasing productivity* measures, making this choice Out of Scope. This choice also does not match with how the term *"Boskinized"* is used in the passage.
B. taking into account technology's role in an improved efficiency.	**(B)** is also Out of Scope because there is no mention of *technology's role* in the passage.
C. reassessing patterns of consumption and quality of product.	This choice is a spot-on match with the prediction, making **(C)** the correct answer.
D. evaluating the inequalities in various levels of incomes.	**(D)** gives us an option that discusses wage inequality. However, this idea was a facet of the author's argument, not of Boskin's revisions. This is a Faulty Use of Detail answer choice.

MCAT Critical Analysis and Reasoning Skills

Question	Analysis
2. The Boskin Report does all of the following EXCEPT:	The word EXCEPT shows us that this is a Scattered Detail question—one that might be worth skipping on Test Day until more time is available. Our Plan is to use the Outline to find relevant details and eliminate them systematically, keeping in mind that the correct answer is the one NOT included in the passage.
A. reveals that the C.P.I. was inaccurate.	Paragraph 3 tells us that the Boskin Report demonstrated that the C.P.I. *had been systematically overstating inflation*, eliminating **(A)**.
B. reconciles the present disparity between productivity and wage levels.	Paragraph 4 shows us that Boskin did reconcile wages and productivity, eliminating **(B)**.
C. reveals the reasons for the increasing disparity between the highest and lowest income earners.	**(C)** is not present in the passage. While the income disparity was discussed, no mention was made of its causes—making this the correct answer.
D. provides possible clarification for economic progress in the 1950s and 1960s.	In paragraph 5, we see that Boskin's work, if applied to the 1950s and 1960s, *could find that official numbers understated the rate of progress*, eliminating **(D)**.

9: Question Types I

Question	Analysis
3. The author mentions the figures in paragraph 6 in order to show that:	The phrase *in order to* shows us that this is a Function question, which usually means fast points. According to our Outline, paragraph 6 focused on how *income inequality increased*.
A. the total productivity of America has not seen a significant increase since the 1970s.	(A) can be eliminated because the focus of paragraph 6 is *income inequality*, not increases in *total productivity*.
B. the income inequality in America is a problem that is not eliminated by revision of the price index.	(B) deals with wage discrepancy and must be the correct answer.
C. each American worker's productivity is directly proportional to overall national productivity gains.	(C) is an Opposite because the author states that *while workers… have shared… in national productivity gains, they have not done so equally*.
D. Boskin's report is unable to explain the discrepancy between productivity growth and wage increases.	(D) deals with a discrepancy, but not the right one. This paragraph focuses on *income inequality*, not *the discrepancy between productivity growth and wage increases*.

Question	Analysis
4. The author's primary purpose in presenting this passage is to:	The words *primary purpose* identify this as a Main Idea question, which usually can be answered quickly. The Goal in this passage is *to argue that the C.P.I., with or without Boskin's work, understated real wages; and that the productivity/wage disparity is more complicated than it first appears.*
A. argue that overreliance on the C.P.I. is insufficient for explaining the current state of the American worker.	**(A)** is correct; the passage primarily addresses the idea that the *C.P.I. ... understated real wages* (and therefore does not fully explain the current state of the American worker).
B. argue that wages actually increased from 1978 to 1994.	In **(B)**, the answer is far too specific as it applies only to paragraph 2 and not the entire passage.
C. argue that a capitalistic oppression of the worker is the primary cause of the current economic climate.	The *capitalist oppression of the worker* is part of the *simplistic* model dismissed in paragraph 7. Because the author does not agree with this model, **(C)** can be eliminated.
D. suggest that partisan division in Congress would be more adequate for explaining the current economic climate.	Congressional divide, while perhaps present in real life, was never mentioned in the passage, making **(D)** Out of Scope.

9: Question Types I

Question	Analysis
5. The author's use of the term "statistical imagination" in paragraph 3 most nearly indicates:	This is a Definition-in-Context question. A quick Scan of paragraph 3 shows us that *statistical imagination* refers to a shortcoming of the traditional C.P.I. model, due to failure to *take account of changes in consumption patterns and product quality improvements*. Be on the lookout for any wrong answers that sound like a standard definition for *imagination*.
A. wage data for the last 25 years has been falsified.	(A) is Out of Scope because we're never told that data was *falsified*.
B. the pessimistic view of the economy indicated by the C.P.I. is overstated due to underestimation of the significance of key variables.	(B) matches closely with the prediction and is the correct answer.
C. mathematical models of the economy are less accurate than anecdotal reports.	(C) is Out of Scope because this passage never compares *mathematical models* to *anecdotal reports* in terms of validity.
D. the C.P.I. is a completely unreliable tool for explaining the economic climate.	The C.P.I. certainly has some issues but is not the *completely unreliable tool* mentioned in (D)—this is too Extreme.

Practice Questions

Passage 1 (Questions 1–6)

The United States has less than half of the 215 million acres of wetlands that existed at the time of European settlement. Wetland conversion began upon the arrival of European immigrants with their traditional antipathy to wetlands and with the will and technology to dry them out. In the mid-19th century, the federal government awarded nearly 65 million acres of wetlands to 15 states in a series of Swamp Land Acts. But the most rapid conversion occurred between the mid-1950s and mid-1970s, when an estimated 450,000 acres per year were lost, primarily to agriculture.

This conversion has meant the loss of a wide range of important wetland functions. Wetlands inhibit downstream flooding, prevent erosion along coasts and rivers, and help remove or assimilate pollutants. They support scores of endangered birds, mammals, amphibians, plants, and fishes. Wetlands provide aesthetic and open-space benefits, and some are critical groundwater exchange areas. These and other public benefits have been lost to agricultural forestry and development enterprises of all kinds, despite the fact that most of the conversion goals might have been obtained with far less wetland loss through regional planning, stronger regulation, and greater public understanding of wetland values.

At best, existing wetland laws and programs only slow the rate of loss. Despite the growing willingness of government to respond, wetland protection faces significant obstacles. Acquisition as a remedy will always be limited by severe budget constraints. The Emergency Wetlands Resources Act allocates only $40 million per year in federal funds, supplemented by relatively modest state funds, for wetland purchase. Ultimately, the wetlands that are protected will be a small percentage of the approximately 95 million acres remaining today. Wetland acquisition by private environmental groups and land trusts adds qualitatively important, but quantitatively limited, protection. Government incentives to induce wetland conservation through private initiatives are limited and poorly funded. Some private developers have recognized that business can protect selected wetlands and still profit. Recreational developments in Florida have benefited from wetland and habitat protection that preserves visual amenities. It is doubtful, however, that these business decisions to save wetlands would have occurred without strong government regulation; the marketplace does not generally recognize the public benefits of wetlands for flood control, fish and wildlife, and other long-term values.

One possible strategy (and the one presently being implemented) is to protect each and every wetland in threatened areas according to stringent permit guidelines that do not distinguish by wetland types or values. This approach may be environmentally desirable, but it has not worked. About 300,000 acres of wetlands are lost each year. An alternative strategy is to develop a regional management approach focused on valuable wetlands in selected areas that are under intense pressure. Broad regional wetland evaluations could identify critical wetland systems that meet particular local and national needs and avoid abandonment of any wetlands without careful review of the tradeoffs. Cooperating federal, state, and local interests can then anticipate and seek ways to prevent wetland losses and can guide future development in areas where alternative options exist. There is no general federal authority to conduct such planning for wetland system protection. But there are several authorities under which a program to anticipate and prevent wetland losses on an area-wide basis can be developed.

1. The author mentions the Emergency Wetlands Resources Act in order to:

   A. emphasize the important role of governmental willingness to preserve wetlands.
   B. prove that federal funding is sufficient for the preservation of wetlands.
   C. advocate passage of new legislation to protect America's wetlands.
   D. show how fiscal constraints affect purchase of wetlands.

2. According to the passage, all of the following contributed to the rapid loss of wetlands in the United States EXCEPT:

   A. technological innovations implemented by European settlers.
   B. development of commercial and residential real estate complexes.
   C. increased rezoning for the purposes of agricultural and industrial operations.
   D. conversion of wetlands for agricultural uses.

3. In the final paragraph, the author's primary purpose is to:

   A. present several potentially effective strategies for protecting wetlands and preventing additional losses of these important conservation areas.
   B. criticize current methods of protecting wetlands for their inefficiency and disregard of strict governmental guidelines.
   C. argue that collaboration between regional, local, and federal organizations is the only way to establish effective wetlands conservation methods.
   D. suggest that the present approach to protection of wetlands has been ineffective and advocate a different method.

4. Why does the author mention federal acquisition of wetlands in the third paragraph?

   A. To suggest a potential remedy to the rapid conversion of threatened wetlands
   B. To emphasize the limitations of current conservation efforts
   C. To show that recreational developments, as seen in Florida, preserve visual amenities
   D. To advocate for increased governmental spending in wetlands conservation

5. Which of the following does the author state as beneficial functions of the wetlands?

   I. Supporting communities of endangered amphibians
   II. Providing renewable forestry options
   III. Erosion and flood control

   A. III only
   B. I and II only
   C. I and III only
   D. I, II, and III

6. The author can best be viewed as a proponent of:

   A. developing new strategies and improving current efforts to prevent wetland losses.
   B. maintaining the current stringent permit guidelines to protect conservation areas.
   C. promoting development of wetlands for visual amenities and aesthetic benefits.
   D. current government incentives to induce wetland conservation through private business funding.

Passage 2 (Questions 7–12)

Although many may argue with my stress on the continuity of the essential traits of American character and religion, few would question the thesis that our business institutions have reflected the constant emphasis in the American value system on individual achievement. From the earliest comments of foreign travelers down to the present, individuals have identified a strong materialistic bent as a characteristic American trait.

The worship of the dollar, the desire to make a profit, the effort to get ahead through the accumulation of possessions, all have been credited to the egalitarian character of the society. As Tocqueville noted in his discussion of the consequences of a democracy's destruction of aristocracy: "They have swept away the privileges of some of their fellow creatures which stood in their way, but they have opened the door to universal competition."

A study of the comments of various 19th-century foreign travelers on American workers reveals that most of these European writers, among whom were a number of socialists, concluded that social and economic democracy in America has an effect contrary to mitigating compensation for social status. American secular and religious values both have facilitated the "triumph of American capitalism" and fostered status striving.

The focus on egalitarianism and individual opportunity has also prevented the emergence of class consciousness among the lower classes. The absence of a socialist or labor party, and the historic weakness of American trade-unionism, appear to attest to the strength of values that depreciated a concern with class.

Although the American labor movement is similar to others in many respects, it differs from those of other stable democracies in ideology, class solidarity, tactics, organizational structure, and patterns of leadership behavior. American unions are more conservative; they are more narrowly self-interested; their tactics are more militant; they are more decentralized in their collective bargaining; and they have more full-time salaried officials, who are on the whole much more highly paid. American unions have also organized a smaller proportion of the labor force than have unions in these other nations.

The growth of a large trade-union movement during the 1930s, together with the greater political involvement of labor organizations in the Democratic party, suggested to some that the day—long predicted by Marxists—was arriving in which the American working class would finally follow in the footsteps of its European brethren. Such changes in the structure of class relations seemed to these observers to reflect stagnancy and limitations on social mobility—the decline of opportunity and the petrifaction of class lines. To them, such changes could not occur without modification of the traditional value system.

A close examination of the character of the American labor movement suggests that it, like American religious institutions, may be perceived as reflecting the basic values of the larger society. Although unions, like all other American institutions, have changed in various ways consistent with the growth of an urban industrial civilization, the essential traits of American trade unions, as of business corporations, may still be derived from key elements in the American value system.

7. In the context of the passage, the phrase "strong materialistic bent," as used in the sentence, "From the earliest comments of foreign travelers down to the present, individuals have identified a strong materialistic bent as a characteristic American trait," (paragraph 1) refers to:

   A. European socialists' view of aristocrats.
   B. European travelers' concern with democracy.
   C. American society's emphasis on acquiring wealth.
   D. American religion's criticism of secular values.

8. Based on the information given in the passage, which of the following is/are NOT true?

    I. American society emphasizes class solidarity over individual achievement.
    II. American unions are less interested in non-labor issues than unions in other democracies.
    III. American labor organizations and American religious institutions share some of the same values.

    A. I only
    B. II only
    C. II and III only
    D. I, II, and III

9. According to the passage, all of the following have influenced the outlook of the American labor movement EXCEPT:

    A. secular values.
    B. religious values.
    C. urban industrial civilization.
    D. foreign labor movements.

10. According to the passage, which of the following is a part of the "traditional value system"?

    A. Class solidarity
    B. Individual achievement
    C. Urban industrialization
    D. Marxist ideology

11. The author of this passage most likely believes that American labor unions:

    A. are more influenced by the American values of competition and materialism than their European counterparts.
    B. sacrifice status and wealth by adhering to religious values.
    C. seemed to reflect more European values during the 1930s than American values.
    D. are stronger, more widespread organizations than European trade unions.

12. Based on its use in the passage, the word "petrifaction" most nearly means:

    A. dissolution.
    B. conversion to stone.
    C. recognition.
    D. solidification.

Explanations to Practice Questions

Passage 1 (Questions 1–6)

Sample Passage Outline

P1. History of wetlands conversion

P2. Benefits of wetlands, ways loss could have been minimized

P3. Current problems with conservation efforts and ways of minimizing further loss

P4. Current solution (save all threatened wetlands) hasn't worked; proposes new regional management strategy

Goal: To describe problems with current wetland conservation efforts and to propose a new strategy

1. D

The phrase *in order to* tells us that this is a Function question. The *Emergency Wetlands Resources Act* is mentioned in paragraph 3, the purpose of which is to present *current problems with conservation efforts and ways of minimizing further loss* according to our Outline. The fact that the paragraph is neutral is sufficient information to determine the answer: *emphasize*, *prove*, and *advocate* are all too strong as purpose verbs to match the author's tone, whereas *show* perfectly matches with his or her neutrality. To ensure that **(D)** is the correct answer, consider that the previous sentence states that buying up the wetlands to protect them *will always be limited by severe budget constraints*. To further explain the incorrect answers, let us examine each individually. While the government's role in preservation is important, the author's focus in the description of the Emergency Wetlands Resources Act is not on emphasizing this role, but instead on explaining why the Act is insufficient to protect the wetlands; therefore, **(A)** is incorrect. The author states that the Emergency Wetlands Resources Act contributes *only* $40 million, indicating that this is not enough and that money is constrained; therefore, **(B)** is incorrect. We can also eliminate **(C)** because the author never discusses *new legislation* in conjunction with the Emergency Wetlands Resources Act; this answer is Out of Scope.

2. C

For this Scattered Detail question, we will have to find three answer choices that are mentioned in the passage and eliminate them. For **(A)**, the *technological innovations* of European settlers are mentioned in paragraph 1. *Development* enterprises are listed at the end of paragraph 2, which eliminates **(B)**. **(C)**, however, cannot be found in the passage. No mention is made of *increased rezoning* of wetlands, so this must be the correct answer. Finally, *conversion of wetlands for agricultur[e]* is listed at the end of the first paragraph, which removes **(D)**.

3. D

The phrase *primary purpose* may make us think this is a Main Idea question, but the fact that it is restricted to *the final paragraph* means that it is a Function question. Our Outline gives us the information we need: the suggested Label for the last paragraph is *Current solution (save all threatened wetlands) hasn't worked; proposes new regional management strategy*. This Label matches **(D)**. **(A)** is flawed because the author presents only one *potentially effective strategy*, not *several*. **(B)** is tempting because the first portion of the paragraph does *criticize current methods*, but this answer omits the main point of the paragraph: the author's proposal. The author does not state that his or her proposal is the *only* way to conserve wetlands, which means **(C)** is Extreme.

4. B

Acquisition is discussed in the third sentence of the paragraph 3. Just like question 1, the Label for paragraph 3 gives enough information to answer the question: the purpose of this paragraph is to present *current problems with conservation efforts and ways of minimizing further loss*. This matches with **(B)**. In this paragraph, the author states that *acquisition as a remedy will always be limited by severe budget constraints*, which is the Opposite of *suggest[ing] a potential remedy*, so we can eliminate **(A)**. **(C)** is a Faulty Use of Detail—*federal acquisition* is not related to *recreational developments*. **(D)** is Out of Scope for this paragraph because the author does not *advocate for increased governmental spending* here.

5. C

The Roman numeral format and the verb *state* tell us that this is a Scattered Detail question. *Beneficial functions of the wetlands* are mentioned in the first part of paragraph 2. Relevant to this question are the details that *wetlands inhibit downstream flooding, prevent erosion along coasts and rivers,...and support scores of endangered birds, mammals, amphibians, plants, and fishes*. Statements I and III are listed in this paragraph, whereas *renewable forestry options*, Statement II, are never mentioned.

6. A

This is a masked Main Idea question; asking what the author is a *proponent of* is another way of asking what the author likes. The author's Goal in the passage, according to the Outline, is to *describe problems with current wetland conservation efforts and to propose a new strategy*. This prediction matches well with **(A)**. In the final paragraph, the author discusses the flaws of the current permit system, so **(B)** can be eliminated as an Opposite. **(C)** can also be eliminated for similar reasons because the author is a proponent of preservation of—not *development of*—the wetlands. Finally, the author states in the third paragraph that *government incentives to induce wetland conservation through private initiatives are limited and poorly funded*, so **(D)** can also be eliminated.

Passage 2 (Questions 7–12)

Sample Passage Outline

P1. Individualism and materialism = American traits

P2. American traits come from egalitarianism, ex: Tocqueville

P3. European travelers' perspective on American system

P4. Egalitarianism and individualism prevent class consciousness

P5. Comparison of American and foreign labor unions

P6. Early predictions of American trade unions by Marxists

P7. American labor movement's values reflect general American values

Goal: To argue that the American labor movement reflects traditional American values

7. C

This is a Definition-in-Context question asking for the author's definition of *strong materialistic bent*. Immediately after the referenced sentence, the author mentions *the worship of the dollar, the desire to make a profit, the effort to get ahead through the accumulation of possessions* as an elaboration of the topic of materialism. Thus, materialism—a characteristically American trait—focuses on acquiring wealth, which matches **(C)**. Because materialism is attributed to Americans, and not *aristocrats* or *democracy* in general, **(A)** and **(B)** can be eliminated. Finally, the author never mentions *criticism* from American religion at all, so we can eliminate **(D)** as Out of Scope.

8. A

This is a Scattered Detail question asking for statements that are *NOT true* based on the passage. Statement I runs counter to the author's main point, most forcefully expressed in the first two paragraphs, that *individual achievement*, *not class solidarity*, is a characteristically American trait. Statement II refers to the differences between American and foreign labor unions, which are listed in paragraph 5. Here, the author argues that American unions are more *narrowly self-interested*—in other words, they focus more specifically on labor issues than foreign unions. Statement II must therefore be true, and the answer must be **(A)**. Statement III is also true based on the beginning of paragraph 7: *the American labor movement,... like American religious institutions, may be perceived as reflecting the basic values of the larger society.*

9. D

This is another Scattered Detail question because of the phrase *According to the passage* and the word *EXCEPT*. Secular and religious values, **(A)** and **(B)**, are addressed in paragraph 3 as both have *facilitated the "triumph of American capitalism."* American capitalism appears to have affected the goals of unions, according to paragraph 5, so these answer choices can be eliminated. *Urban industrial civilization* is listed as influencing unions in paragraph 7, so we can eliminate **(C)**. Foreign labor movements contrasted with the American labor movement in paragraph 5, but no influence is described here. The closest match in the passage would be the Marxist prediction that *the American working class would finally follow in the footsteps of its European brethren*, but this prediction actually failed to pan out. Therefore, **(D)** is correct.

10. B

For this Detail question, we can look for something that fits in with the two values the author mentioned as part of the traditional American value system: individual achievement and egalitarianism. **(B)** matches perfectly and is the correct answer. **(A)** and **(D)** are Opposites because *class solidarity* and *Marxist ideology* are more likely considered values of foreign democracies according to paragraph 5. Finally, while **(C)** is listed in the final paragraph as influencing labor movements, it is not part of the American traditional value system as described throughout the passage, so it can be eliminated.

11. A

This is an Inference question of the Implication subtype. Here, we should look to our Passage Outline and the author's Goal to determine his or her opinion. The author makes the case that American labor unions are more strongly influenced by traditional American values, such as individual achievement and egalitarianism, rather than more European values. This prediction is very close to **(A)**; *competition* and *materialism* are also mentioned as American characteristics in paragraphs 2 and 1, respectively. **(C)** is the Opposite of this prediction, so it can be eliminated. **(B)** is Out of Scope because *sacrific[ing] status and wealth* is not discussed as a religious value here. Finally, **(D)** is also Out of Scope; the author never talks about the relative *strength* of American labor unions as compared to European trade unions.

12. D

This is another Definition-in-Context question, so let's locate the relevant sentence: *Such changes in the structure of class relations seemed...to reflect stagnancy and limitations on social mobility—the decline of opportunity and the petrifaction of class lines.* There's parallelism here in the sentence's construction: *stagnancy* refers to *the decline of opportunity*, whereas *limitations on social mobility* refers to *the petrifaction of class lines*. Putting this together, if social mobility is limited, then class lines must be harder to break through, which matches with **(D)**. **(A)** is an Opposite, as *dissolution* of class lines would enable social mobility. Mere *recognition* of these lines, as in **(C)**, would not necessarily limit social mobility, thus eliminating this answer choice. Finally, **(B)** is far too literal—*petrifaction* can also mean the conversion of organic matter into stone, but an intangible concept like *class lines* would not actually be turned into stone.

10

Question Types II: *Reasoning Within the Text* Questions

10: Question Types II

In This Chapter

10.1 Inference Questions — **230**
- Sample Question Stems — 230
- Strategy — 231
- Worked Example—
  - A History Passage — 233

10.2 Strengthen–Weaken (Within the Passage) Questions — **236**
- Sample Question Stems — 236
- Strategy — 237
- Worked Example—
  - An Ethics Passage — 238

10.3 Other *Reasoning Within the Text* Questions — **242**
- Clarification — 242
- Weakness — 242
- Paradox — 243

Concept and Strategy Summary — **244**

Worked Example — **246**

Introduction

> **LEARNING GOALS**
>
> After Chapter 10, you will be able to:
>
> - Identify Inference, Strengthen–Weaken, and Other *Reasoning Within the Text* questions
> - Solve *Reasoning Within the Text* questions with focused strategies
> - Apply the major principles of argument and logical structure to MCAT questions

How many times have you been told that *you should never assume*? Whether you were simply told that assumptions breed mistakes or were given some more colorful rationale, this generally sound advice can go only so far. In reality, it is impossible *not* to make assumptions because an assumption is just anything that we take for granted—that we believe without additional evidence. It simply does not matter how rational or scientifically curious you consider yourself to be: at some point you stop asking *Why?* and start to take things on faith. Authors of passages used in *Critical Analysis and Reasoning Skills* (CARS) are of course no different, and the Association of American Medical Colleges (AAMC) will test that you can identify the assumptions that they hold, as well as the other essential elements of the reasoning they employ in their passages.

MCAT Critical Analysis and Reasoning Skills

In this chapter, we'll continue the treatment of question types first employed with *Foundations of Comprehension*, now examining the two types of *Reasoning Within the Text* questions that Kaplan has identified: Inference questions, which ask about the unstated parts of arguments, and Strengthen–Weaken (Within the Passage) questions, which predominantly concern the ways in which arguments are backed by evidence and undermined by refutations. We adopt the same general approach as before: after identifying what makes the questions distinctive and offering several common question stems, we discuss strategies for the tasks each type involves, illustrating them with a few worked examples.

Reasoning Within the Text questions account for approximately 30 percent of what you'll encounter on Test Day, according to both the AAMC's official statements and Kaplan's own extensive research of released AAMC material. Our data indicate that a bit more than half fall into the Inference category, so this type will account for approximately one-sixth of the CARS section or about eight or nine questions. This means Inference questions are roughly as common as Detail and Strengthen–Weaken (Beyond the Passage) questions, but they are less common than Apply questions. Strengthen–Weaken (Within the Passage) questions are slightly rarer, averaging about six questions per section.

Note: The question types, as well as the Kaplan Method for CARS Passages, Kaplan Method for CARS Questions, and Wrong Answer Pathologies are included as tear-out sheets in the back of this book.

10.1 Inference Questions

What makes **Inference questions** distinctive is that they deal with unstated parts of arguments: information that is not explicitly written by the author but that *must* be true given what is claimed in the passage. It is crucial to understand that the correct answers to Inference questions are not simply assertions that are *possibly* true or that *could* be accurate; rather, they are necessary assumptions or at least highly probable implications. We call these kinds of questions Inference because, as first noted in Chapter 5 of *MCAT CARS Review*, it is the common name of the process used to arrive at both **assumptions** (unstated evidence) and **implications** (unstated conclusions), collectively known as inferences.

MCAT Expertise

According to our research of released AAMC material, Inference questions make up about 16 percent of the CARS section (about eight or nine questions).

SAMPLE QUESTION STEMS

- In [paragraph reference], it is reasonable to believe that the author assumes:
- Implicit in the discussion of [topic] is the underlying assumption that:
- The passage suggests [claim] because:

10: Question Types II

- On the basis of the author's account of [topic], which of the following might reasonably be inferred?
- The author implies that [concept] is NOT:
- It is reasonable to conclude that the author regards [person or position] as:
- The author says [quotation], but also [paraphrase of different claim]. These beliefs imply:
- Which of the following inferences is most justified by information provided in the passage? [list of Roman numerals]

As these examples indicate, during the first step of the Kaplan Method for CARS Questions—in which you Assess the question type—you can recognize that you might be dealing with an Inference question if you see one or more of the following words (or variations of them): *assume*, *because*, *conclude*, *imply*, *infer*, *justify*, *reasonable*, or *suggest*. However, if a question stem contains any of these words and new information of some kind, then it will fall into the *Reasoning Beyond the Text* category instead. If there are no new elements, you are likely dealing with an Inference question.

There is actually something of a continuum between Detail and Inference questions; the position of questions on that continuum varies based on the complexity of the reasoning used to solve them. A question that tells you to make an *inference* might require one relatively simple step, such as canceling out a double negative or identifying a paraphrase of lines from the passage. Sometimes you'll even find yourself looking for an item that the *passage suggests* or the *author implies* but, after Executing your Plan, discover that the answer was in the text virtually word for word. In cases such as these, consider yourself fortunate because you've uncovered a Detail question in disguise! The predictions you make for these questions will likely be more than adequate to reach the correct answer.

The downside to this ambiguity is that sometimes a question that uses simple declarative language will require a lot more critical thinking than it initially seems. Just because a stem includes a phrase like *the author asserts* or *as stated in the passage* (common question stems for Detail questions) does not necessarily mean that the answer is stated straightforwardly. Notwithstanding such trickery on the part of the AAMC, as long as you recognize the possibility that apparent Detail questions can be disguised as Inference questions and are prepared to apply the Denial Test strategy to those questions, you can avoid being fooled!

STRATEGY

There are only two basic tasks with the Inference question type. When you recognize that you need to make an inference, first ask yourself whether you are looking for a missing but essential piece of evidence (an **assumption**) or for a conclusion that is unstated but highly probable given what is said (an **implication**).

Assumptions

If your task is to identify an assumption, the first prong of your attack Plan should be to determine what claim it is supposed to underlie. Sometimes the question stem will say this explicitly, or it may use quoted text or paragraph numbers to refer to a particular claim. Use your Outline if necessary to find the relevant sentences from the passage. When you Execute your Plan, reread those sentences and isolate the specific statement, taking care to see whether there's any existing evidence in the surrounding text that's used to support it. Logic keywords are your best friends here, but keep in mind that a lot of authors use them sparsely, so they won't always be there to help. The Answer to assumption questions might include words or ideas that are similar to a piece of evidence actually stated, so you can use any that you locate for your prediction, or just go with whatever links the evidence provided to the conclusion when you reread the text—*now I see what the author is taking for granted!*

Implications

Whenever the task is finding an implication, isolate in the text any particular sentences referred to in the question stem and watch out for Logic keywords, just as with assumptions. Now, however, you'll be looking to see whether the particular statements referenced are used to support anything else. If so, use those explicitly supported conclusions to set your expectations for the correct answer because sometimes implications will be quite similar to them. Alternatively, if another implication occurs to you when you reread, you can use that as a prediction.

Whenever you can't find a match for your prediction—if you thought, at first, it was a Detail, if you didn't know where to look because the stem had no paragraph reference clues, or if the answer choices just turned out very differently than you expected—Plan B is to use a special version of process of elimination known as the **Denial Test**. Let's take a look at how it works in practice.

The Denial Test

For each answer choice, negate (take the opposite of) what is being said. In question stems that end with a colon (:), you may need to take part of the text from the end of the stem and combine it with the answer choice to create a sentence that can then be denied. Be careful with sentences that already contain negative words, prefixes, and suffixes because sometimes just removing that text is not enough to change the meaning in the proper way, as shown in the first worked example later in this section. If nothing else, any claim can be denied by adding *It is not the case that* to the beginning of the sentence. Once you've denied the claim, think about the effect it has on the passage. If you're not sure what effect it has, look for clues in the question stem or the choice itself to see if it refers to

> **Key Concept**
>
> Inferences are unstated parts of arguments that must be true based on what the passage says. Assumptions are unstated pieces of evidence, while implications are unstated conclusions.

10: Question Types II

a particular part of the passage that you could reread to get an idea. Don't forget to use your Outline to help navigate. If it's clear that the negated claim has no significant impact, then cross it off.

When you come across an answer choice that logically conflicts with the text once you've negated it, you've likely found the correct answer. However, sometimes multiple answer choices will have denials that cause problems for the text, so when Executing the Denial Test you should *always test every answer choice* and pick the one for which negation has the most detrimental impact on arguments or assertions from the passage. Keep in mind that this can be time-consuming, which is why it's generally a good idea to triage the question once you discover no quicker approach will serve.

It is hard to understand how the Denial Test works simply by reading about it. Check out the worked example below and make sure to practice the Denial Test on Inference questions to get used to using it.

> **MCAT Expertise**
> While the Denial Test will always reveal the correct answer in an Inference question, it's very time-consuming. If you cannot set good expectations for the right answer during the Execute step, triage the question, and return to it later with the Denial Test.

WORKED EXAMPLE—A HISTORY PASSAGE

In 1941, an exuberant nationalist wrote: "We must accept wholeheartedly our duty and our opportunity as the most powerful and vital nation...to exert upon the world the full impact of our influence, for such purposes as we see fit and by such means as we see fit." If forced to guess the identity of the writer, many US citizens would likely suspect a German jingoist advocating for *Lebensraum*. In actuality, the sentiment was expressed by one of America's own: Henry Luce, the highly influential publisher of the magazines *Life*, *Time*, and *Fortune*. Luce sought to dub the 1900s the "American Century," calling upon the nation to pursue global hegemony as it slipped from the grasp of warring Old World empires. As a forecast of world history, Luce's pronouncement seems prescient—but is it justifiable as a normative stance?

Not all of Luce's contemporaries bought into his exceptionalist creed. Only a year later, Henry Wallace, vice president under FDR, insisted that no country had the "right to exploit other nations" and that "military [and] economic imperialism" were invariably immoral. It is a foundational assumption in ethics that the wrongness of an act is independent of the particular identity of the actor—individuals who pay no heed to moral consistency are justly condemned as hypocrites. So why should it be any different for nation-states? In accord with this principle, Wallace proselytized for "the century of the common man," for the furtherance of a "great revolution of the people," and for bringing justice and prosperity to all persons irrespective of accidents of birth. Sadly, Wallace never had the

MCAT Critical Analysis and Reasoning Skills

chance to lead the United States in this cosmopolitan direction; prior to Roosevelt's demise at the beginning of his fourth term, the vice presidency was handed to Harry Truman, a man whose narrow provincialism ensconced him firmly in Luce's camp. And with Truman came the ghastly atomic eradication of two Japanese cities, the dangerous precedent set by military action without congressional approval in Korea, and a Cold War with the Soviet Union that brought the world to the brink of nuclear destruction.

> **Example:**
>
> 1. One can most justifiably conclude on the basis of the author's discussion that Henry Luce assumed that:
>   A. the United States did not have the right to create a military or economic empire.
>   B. nation-states are never bound by the same ethical principles that persons are.
>   C. the same normative standards should apply to both Americans and Germans.
>   D. moral rules that govern individual behavior do not necessarily apply to countries.
>
> **Solution:** Assessing that the question is asking for an assumption is as easy as reading *assumed* at the end of the stem. You'll want to Plan to use the Denial Test, but first be clear on what Henry Luce's position is, particularly because he's not the only *Henry* discussed in the passage. Luce advocated for what the author calls the *exceptionalist* stance that the United States, as the most powerful country at the time, was free to do as it pleased. This is in contrast to Henry Wallace, who explicitly rebuffed Luce's view, so watch out for Faulty Uses of Detail that would actually describe Wallace.
>
> Denying **(A)** yields *the United States **did** have the right to create a military or economic empire*, which is completely consistent with Luce's view. Eliminate it. As written, **(A)** is actually a view attributed to Wallace by the author, so it's both a FUD and an Opposite because it was the major point of disagreement between the two.
>
> **(B)** contains *never*, which is an Extreme word, but because Luce seems to have a fairly strong position, you don't want to jump to Labeling it a Distortion quite yet. It's a bit trickier to negate: to say that it is *not* the case that some event *never* happens is the same as saying the event *sometimes* occurs. Thus, the proper negation of **(B)** is *nation-states are **sometimes** bounded by the same ethical principles that persons are*, which is completely consistent with Luce, who presumably believes that ethical principles do apply to nations sometimes, say on those occasions when the nation in question is not the United States. Eliminate it.

Bridge

Remember that the opposite of an Extreme Negative keyword (like *never*) is a Moderating Positive keyword (like *sometimes*). The circular view of Author keywords presented in Chapter 3 of *MCAT CARS Review* is a great way to visualize this change.

For **(C)**, the contradiction would be *the same normative standards should **not** apply to both Americans and Germans*, which is again consistent with Luce's *"American Century"* idea. Hence, we can definitely cross off **(C)**, which, as written, is actually another one of Wallace's beliefs—a second Opposite/FUD combo.

All that remains is **(D)**, which does indeed destroy the argument if rejected. Countering it by saying *moral rules that govern individual behavior **do** necessarily apply to countries* would mean that they *always* apply. But this supports the antihypocrisy argument that the author makes in favor of Wallace and against Luce at the beginning of paragraph 2. Thus, denying **(D)** would make Henry Luce's argument fall apart, and so it is an assumption Luce has made. Discarded **(B)** was simply a more extreme version of this claim, rightly rejected as a Distortion because an author is more likely to assume a weak form of a statement than a strong one.

Example:

2. It is reasonable to infer that the author believes that:
   A. Harry Truman was the worst US president of the 20th century.
   B. Franklin Roosevelt did not endorse the idea of the "American Century."
   C. Henry Wallace would not have approved of the use of atomic weapons.
   D. Henry Luce did not provide an accurate historical assessment of the 1900s.

Solution: We can easily Assess this question as an Inference question given the phrase *reasonable to infer*. With a stem like this, which has no specific clues or references, you have little choice but to proceed with the Denial Test. It's not entirely clear what kind of inference this is because an author believes both assumptions and implications, but the same Plan can work for either, as noted above. While you can guess based on the author's descriptive language that he or she does not approve of Harry Truman, you do not know the writer's feelings toward any of the other American presidents in the 1900s, so denying that Truman was the worst president would not have too much of an impact. **(A)** is a Distortion, too Extreme to attribute to the author.

MCAT Critical Analysis and Reasoning Skills

For **(B)**, you do not really know what the author believes about Franklin Delano Roosevelt (FDR). We only know that he was the US president serving over both Wallace and Truman. Therefore we can make no safe inferences about FDR's attitude toward Luce's point of view: this Out of Scope option should be discarded.

Negate **(C)** and you arrive at *Henry Wallace **would** have approved of the use of atomic weapons*. If this were so, had he become president, he would have been guilty of the very misdeeds for which the author blames Truman—and the author would be utterly inconsistent in praising the one and condemning the other. In fact, because the author rails against hypocrisy, it is clear that he or she would have to hold the two men to the same ethical standards. Thus, denying **(C)** would considerably undermine the author's argument in paragraph 2, and this is almost certainly the answer.

To make sure, finish Executing the Denial Test: rejecting **(D)** would have no negative impact—in fact, it would support the author's claim that Luce's view *seem[ed] prescient* as an historical prediction. This assumption suggests that the author indeed views Luce's view as historically accurate, even if ethically questionable. You can now be confident that **(C)** is the Answer.

10.2 Strengthen–Weaken (Within the Passage) Questions

MCAT Expertise

According to our research of released AAMC material, Strengthen–Weaken (Within the Passage) questions make up about 12 percent of questions in the CARS section (about six questions).

Strengthen–Weaken questions span two of AAMC's delineated categories, but both types generally concern the logical relationships between conclusions and the evidence that *strengthens* them or the refutations that *weaken* them. Recall that these relationships were discussed in depth in Chapters 5 and 6 of *MCAT CARS Review*, so return there if necessary. Note that the only substantial difference between Strengthen–Weaken (Within the Passage) and Strengthen–Weaken (Beyond the Passage) is that the former stick to the passage as written while the latter will bring in some new element, usually appearing in the question stem, though on occasion only in the answer choices.

SAMPLE QUESTION STEMS

- The author's suggestion that [claim] is supported in the passage by:
- For which of the following statements from the passage does the author provide the most support?
- The author states in [paragraph reference] that [claim]. This most strengthens the author's contention that:

- Which of the following objections considered in the passage most WEAKENS the author's thesis?
- Which of the following is a claim that the author makes without providing evidence?
- The view of [person] is challenged in the passage by:
- How does other information from the passage relate to the claim that [quotation]?
- What significance does the assertion that [claim] have for the author's argument?

As this list suggests, these types of questions often contain references indicating that the answers will be taken directly from the text and are heavy on words indicating connections between claims like *relate*, *support*, and *challenge*. Some question stems may be ambiguous about whether the support or challenge you're looking for will be coming from in the text or outside of it, in which case you'll also want to bear in mind the strategy for Strengthen–Weaken (Beyond the Passage) questions, detailed in the next chapter.

STRATEGY

Every Strengthen–Weaken question has three pieces: **two claims** and the **connection** between them. You will always be given at least one of these elements, and your task will be to find the other(s), so begin your Plan step by identifying where each piece can be found: either directly in the stem itself, somewhere in the passage, or in the answer choices.

If the *connection* is revealed in the question stem, it will typically be some variation of strengthen (support) or weaken (challenge), as the name of these questions suggest. However, when the connection does not occur until the answer choices, such as when a stem uses vague words like *relevance*, *significance*, or *impact*, the claims occasionally prove to have some other relationship, such as identity (meaning the same thing) or even irrelevance. Once you know whether your task is to Strengthen, Weaken, or find some yet-to-be-discovered relevance, the next step of your Plan is to research the status of any *claims* quoted or otherwise referenced in the question stem. Your Execution could vary based on the number of claims in the stem.

If no other claims are mentioned, such as in a question like *Which of the following passage assertions is the LEAST supported?*, you should probably just decide to triage after your initial Assessment of the question, resorting to process of elimination when you have time to Execute a Plan *later* in the section.

MCAT Critical Analysis and Reasoning Skills

Key Concept
Evidence is used to support a conclusion through a one-way relationship. A refutation is used as a counterargument against a conclusion through a one-way relationship.

If the question stem refers to two claims then the task must be to find the nature of the relationship they share, so think about whether one claim stands under the other, remembering that evidence makes a conclusion more likely to be true and refutations make conclusions less probable, as explained in Chapter 5 of *MCAT CARS Review*.

In most cases, though, you'll just be presented with one claim in the question stem, so Plan to identify the given statement as a conclusion, piece of evidence, or a refutation and then go to the passage, starting with the relevant sentence and looking in the surrounding text for language that suggests the appropriate relationship. When you Execute the Plan, Logic keywords are just about as important for Strengthen–Weaken questions as they are for Inference questions. If the question stem specified the relationship, pay special attention to that one; otherwise, keep an eye out for any logical connections made to the claim, using those to set expectations.

If your initial Plan of attack proves unsuccessful, try process of elimination, crossing out any answer choice that does not establish the correct kind of relationship. Do not forget that support is unidirectional: if the "arrow" points the wrong way, it cannot be the right choice. So, for instance, if you are asked to find a claim that *supports* the author's thesis, a potential wrong answer is an implication that could be drawn if you assumed the thesis was true—in other words, a conclusion that the thesis itself *supported*.

WORKED EXAMPLE—AN ETHICS PASSAGE

The most prevalent argument against doctor-assisted suicide relies upon a distinction between *passive* and *active* euthanasia—in essence, the difference between killing someone and letting that person die. On this account, a physician is restricted by her Hippocratic oath to do no harm and thus cannot act in ways that would inflict the ultimate harm, death. In contrast, failing to resuscitate an individual who is dying is permitted because this would be only an instance of refraining from help and not a willful cause of harm. The common objection to this distinction, that it is vague and therefore difficult to apply, does not carry much weight. After all, applying ethical principles of *any sort* to the complexities of the world is an enterprise fraught with imprecision.

Rather, the fundamental problem with the distinction is that it is not an ethically relevant one, readily apparent in the following thought experiment. Imagine a terminally ill patient hooked up to an unusual sort of life support device, one that only functioned to prevent a separate "suicide machine" from administering a lethal injection so long as the doctor pressed a button on it once per day. Would there be any relevant difference between using the suicide machine directly and not using the prevention device? The intention of the doctor would be the same (fulfilling the patient's wish to die), and the effect would be the same (an injection causing the patient's death). The only variance here is the means by which the effect comes about, and this is not an ethical difference but merely a technical one.

> **Example:**
>
> 3. Which of the following roles is played in the passage by the claim that the difference between killing and letting die is ethically relevant?
>   I. It is contradicted by the assertion that the distinction between active and passive euthanasia is only technical.
>   II. It bolsters the contention that applying ethical principles precisely is difficult.
>   III. It underlies the most common argument against physician-assisted suicide.
>   A. III only
>   B. I and II only
>   C. I and III only
>   D. I, II, and III
>
> **Solution:** The question asks about *roles* that are *played in the passage* by a statement given in the stem, so you can Assess that this is a Strengthen–Weaken (Within) question. However, it's a Roman numeral question, and these are usually more time-consuming. Your best bet is to save this for the end of the question set.
>
> When you do work on it, Plan to start by locating the claim referenced in the question and leave the Roman numerals aside for now. As is often the case in more complex Strengthen–Weaken (Within) questions, there is no single sentence that contains all the words in the assertion; rather, parts of it are spread throughout the text. The two lines that are most important are the opening sentences of each paragraph, the first of which refers to *the difference between killing someone and letting that person die* and the second of which maintains that *the distinction…is not an ethically relevant one*.
>
> Just from this analysis, you can see first that the claim that the distinction is ethically relevant must be what is *relie[d] upon* (Evidence keyword) by the so-called *most prevalent argument against doctor-assisted suicide*. In other words, the claim that the distinction is ethically relevant plays a supporting role in that argument. Second, it is clear that the second paragraph is denying this claim. These initial observations already offer a sense of two roles that the distinction plays, a fairly thorough prediction.

MCAT Critical Analysis and Reasoning Skills

MCAT Expertise

CARS authors often use multiple *terms* (words or phrases) to describe the same *concept*, or underlying idea. On Test Day, pay special attention to the ways in which authors use terms, especially when you see Opposition keywords, as discussed in Chapter 3 of *MCAT CARS Review*. When you come across dualisms, you can draw a set of columns adjacent to your map on your noteboard and jot down what words the author uses for each side of the contrast. This can serve as a handy reference for any synonymous language you might encounter in the questions and answer choices. So, for the ethics passage, you would put *killing* and *active euthanasia* in one column and *letting die* and *passive euthanasia* in the other.

At this point, you can look at the answer choices to see how the Roman numerals are distributed. We generally recommend starting with the most common numeral or, alternatively, whichever seems easiest for you. Statements I and III both appear three times, so start with the shorter of them. Statement III suggests the claim *underlies* the most common argument, which is precisely as predicted. Therefore, Statement III must be true and **(B)** can be crossed off.

Turning to Statement I, you'll note the mention of *the distinction between active and passive euthanasia*, which you were told in the first sentence was, *in essence, the difference between killing someone and letting that person die*. This is consistent with the expectation set earlier that the second paragraph challenges the assertion that the claim is ethically relevant. The final sentence confirms it: *this is not an ethical difference but merely a technical one*. The *not* tells you that this is the contradiction that Statement I suggests, so it must also be true, eliminating **(A)**.

There are still two answer choices remaining, so you will have to deal with Statement II. The contention that it mentions did not figure into our prediction, so check the text to find the reference, which is located at the end of the first paragraph. How does this assertion relate to the original claim that the distinction is ethically relevant? The clue is the keyword that precedes the assertion: *After all*. Even though it may sound like a Conclusion keyword on the surface, it's actually an Evidence keyword, which means that this assertion about *applying ethical principles* is in truth used to support something else. Specifically, this statement bolsters the author's belief that the *common objection* to the distinction carries little weight. This is not the relationship suggested by Statement II, which says that the claim that the distinction is ethically relevant supports how hard it is to apply ethical principles precisely. Thus, Statement II is false. Only Statements I and III are true, making **(C)** the correct answer.

Example:

4. On the basis of the author's discussion, which of the following items from the passage LEAST challenges the argument for the prohibition of active euthanasia?
   A. The thought experiment involving two suicide machines from the second paragraph
   B. The assertion that the distinction between passive and active euthanasia is too difficult to apply
   C. The argument that the distinction between passive and active euthanasia is only technical
   D. The claim that the effect and the intention are the same regardless of the type of euthanasia

Solution: Question 4 is somewhat tricky to untangle, with its multiple negative terms, but you can quickly Assess that *challenges* and *from the passage* reveal this to be a Strengthen–Weaken (Within) question—specifically, Weaken. However, the *LEAST* means a **Scattered** format, one in which you'll probably have to test all of the answer choices. Save this for *later* if possible.

When Planning, begin by clarifying the *argument for the prohibition of active euthanasia*. Although the order after the dash in the first sentence is switched, it should be clear from the subsequent sentence that *active euthanasia* refers to the act of *killing*, which is supposedly forbidden because of the Hippocratic oath that the physician takes. This argument prohibiting active euthanasia is in fact that *most prevalent argument* from the first line. The correct answer, then, will be the one that challenges this argument the least.

The Scattered form suggests process of elimination, but before resorting to that, it doesn't hurt to see whether the author actually talks about a challenge that he or she regards poorly because that could be the very answer you're seeking. Indeed, the author does say in paragraph 1 that *the common objection…does not carry much weight*. The objection referenced is that it's difficult to apply the distinction between killing and letting die, so this can serve as your prediction.

Looking at the answers, you can see that this prediction matches **(B)**. However, you should be cautious with this sort of question. While the passage says that the objection carries little weight, it does not say that it carries no weight at all, meaning that if there were an answer choice that had no effect or even supported the argument, that would be *even less* of a challenge. As it turns out, the remaining answer choices are all aspects of the counterargument made in the second paragraph, and all do indeed challenge the original argument. Now you can be confident that **(B)** is the correct answer.

MCAT Critical Analysis and Reasoning Skills

10.3 Other *Reasoning Within the Text* Questions

There are a few rarer types of questions that do not neatly fall into either the Inference or Strengthen–Weaken (Within) categories, but that definitely concern passage reasoning and that do show up in some CARS sections. These can take many different forms and all are rare, so we'll just focus our discussion on three typical tasks.

CLARIFICATION

Questions that ask about **clarification** concern a relationship that is very similar to support, as it is also a one-way relationship. One assertion clarifies another if the two share roughly the same meaning, but the "clarifying" part is typically more specific or exact. Because the clarifying language tends to be more precise, its truth value is easier to assess, and thus you should think of "clarifying" statements as supporting evidence for "clarified" conclusions. Approach them more or less as you would a Strengthen–Weaken (Within) question, except keep in mind that the meanings should be roughly synonymous.

> **Key Concept**
>
> In a Clarification question, look for an answer choice that is nearly synonymous with the given claim, only that is more specific or exact.

With the ethics passage, for instance, you could see a Clarification question like *Which of the following clarifies the author's statement that the common argument against physician-assisted suicide rests upon the distinction between passive and active euthanasia?* The answer would most likely come from one of the sentences that followed, which explained the difference between the two more concretely, including the reasons why one is supposedly permitted and the other is not. In addition to words like *clarify*, words like *explain* and *reflect* are used in questions to indicate this kind of relationship.

WEAKNESS

Weakness questions are somewhat related to Inference questions, but they concern *implicit weaknesses* and *reasonable objections* to arguments discussed in the passage. Instead of the Denial Test, the best Plan is process of elimination by directly assessing the effect that answer choices have on the argument in question. The correct answer will have the most significant negative impact on the argument, perhaps even contradicting it altogether.

> **Key Concept**
>
> Answering a Weakness question is just like using the Denial Test, discussed earlier for Inference questions. The difference is that the correct answer choice will be detrimental to the arguments in the passage *without* being negated.

One example of a Weakness question for the ethics passage above would be *Which of the following is the greatest inherent weakness in the author's use of a thought experiment to support the main argument?* This is a more complex type of *Reasoning Within the Text*, and it is one among a number of rarer questions that require you to appraise the strength of the author's reasoning. The answer to this example might be the fact that thought experiments force the author to rely upon readers' imagination and intuition, which may not always result in the same conclusion as the author intended.

PARADOX

Finally, by a **paradox**, we mean an *apparent* logical contradiction, a set typically consisting of two assertions that seem inconsistent, but only at first glance. These will usually include two distinct claims from the text, phrased in a way to make them sound conflicting, followed by a question like *How would the author resolve this dilemma?* or *How might the passage account for this discrepancy?* Sometimes one of the claims will be a new element, which would technically make such questions *Reasoning Beyond the Text*, although they should still be approached with the same strategy in this case.

The correct answer to a Paradox question must be *consistent* with both of the claims given in the question stem. If possible, it should also not conflict with anything that the author says elsewhere in the passage. Thus, to resolve paradoxes, you should use process of elimination, marking out any answer choice that is inconsistent with one or both of the claims (or with the passage as a whole).

> **Key Concept**
> A paradox is a set of two claims that appear to be inconsistent on the surface. The correct answer in a Paradox question will be consistent with both of the claims, and it will usually attempt to explain the surface inconsistencies between the two claims.

Conclusion

Although often variable in appearance, *Reasoning Within the Text* questions test only a few essential skills: identifying necessary assumptions and inherent weaknesses in arguments; drawing probable inferences from stated claims; understanding relationships of consistency and conflict; and recognizing the connections in passages between conclusions, evidence, and refutations. Regardless of how challenging these questions may seem to you now, you have the ability to improve your reasoning skills! The solution is to practice using Logic keywords to identify support relationships, applying the Denial Test when appropriate, and applying the other strategies discussed in this chapter to your Plan of attack. These tactics will also be useful when working on the final class of questions, *Reasoning Beyond the Text*, the subject of the upcoming chapter.

MCAT Critical Analysis and Reasoning Skills

CONCEPT AND STRATEGY SUMMARY

Inference Questions

- Assess: **Inference questions** look for unstated parts of arguments.
    - Unstated parts of arguments *must* be true given what is claimed in the passage.
    - **Assumptions** are unstated evidence.
    - **Implications** are unstated conclusions.
    - These questions often contain words like *assume*, *because*, *conclude*, *imply*, *infer*, *justify*, *reasonable*, or *suggest*.
- Plan: Determine whether you are looking for an assumption (evidence) or implication (conclusion). Then determine which claim the answer is supposed to support (assumptions) or be supported by (implications).
- Execute: Reread the relevant sentence, noting the explicit evidence and conclusions given.
    - For assumption questions, the answer is either similar to the evidence given or links the evidence to the conclusions.
    - For implication questions, the answer is either similar to the conclusions given or is another logical conclusion one could draw from the evidence.
- Answer: Match your expectations with the right answer. If there is no clear match, or if you cannot perform any of the earlier steps of the Kaplan Method for CARS Questions, use a special form of process of elimination called the Denial Test:
    - Negate each answer choice.
    - Whichever answer choice—when negated—has the most detrimental effect on the argument made in the passage is the correct answer choice.

Strengthen–Weaken (Within the Passage) Questions

- Assess: **Strengthen–Weaken (Within the Passage) questions** concern the logical relationship between conclusions and the evidence that strengthens them or the refutations that weaken them.
    - These questions often contain words like *relate*, *support*, *challenge*, *relevance*, *significance*, or *impact*.
    - These questions are closely related to Strengthen–Weaken (Beyond the Passage) questions, which simply bring in a new piece of information rather than using information directly from the passage.

244

- Plan: Determine the two claims and the connection between them; you will usually be given at least one of these elements and will have to find the other(s).
    - Identify where each piece of the argument can be found: in the question stem, in the passage, or in the answer choices.
    - If no claims are given in the question stem, Plan to triage it and answer it by process of elimination later.
    - If one claim is given in the question stem, determine if it is a conclusion, a piece of evidence, or a refutation.
    - If two claims are given in the question stem, identify the relationship between them.
- Execute: Research the relevant text to determine the missing claim or the connection between the claims. Use Logic keywords to help assemble the argument.
- Answer: Match your expectations with the right answer. If there is no clear match, or if you cannot perform any of the earlier steps of the Kaplan Method for CARS Questions, use process of elimination.

Other *Reasoning Within the Text* Questions

- **Clarification questions** ask for statements that are roughly synonymous, but the clarifying statement tends to be supporting evidence for the conclusion because it is more specific or exact.
    - These questions often contain words like *clarify*, *explain*, or *reflect*.
    - Approach these questions as you would Strengthen–Weaken (Within) questions, except that the meanings of the two claims should be roughly synonymous.
- **Weakness questions** ask for implicit refutations to arguments discussed in the passage.
    - These questions often contain words like *implicit weaknesses* or *reasonable objections*.
    - Approach these questions using the Denial Test for Inference questions, except that the correct answer will be the most detrimental to the argument made in the passage *without* being negated.
- **Paradox questions** ask for the resolution of an apparent logical contradiction.
    - These questions often contain words like *paradox*, *dilemma*, or *discrepancy*.
    - Approach these questions through the process of elimination, crossing out any answer choice that is inconsistent with one or both of the claims of the paradox or with the passage as a whole.

MCAT Critical Analysis and Reasoning Skills

WORKED EXAMPLE

Use the Worked Example below, in tandem with the subsequent practice passages, to internalize and apply the strategies described in this chapter. The Worked Example matches the specifications and style of a typical MCAT *Critical Analysis and Reasoning Skills* (CARS) passage.

Passage	Analysis
Certain contemporary forms of literary criticism draw on modern sociology and political science to understand literary works. There has been a conservative reaction to these schools of criticism, accusing them of imposing modern ideas on old texts. For example, some would consider it an implausible claim that Shakespeare's *The Tempest* can be interpreted as a play about "colonialism" and "imperialism"; after all, these terms were not even in use when Shakespeare wrote the play. These concepts must therefore be modern ones, and it is anachronistic to suppose that Shakespeare had them in mind. Besides, as Ben Jonson wrote, Shakespeare "was not of an age, but for all time," and it trivializes his genius to suppose that he had in mind the fashionable concerns of any one period. The conservative reading of *The Tempest* sees it as a play about "universal" themes like estrangement and reconciliation.	The first words of the paragraph are *certain... forms of literary criticism*, implying that later these forms of criticism will be contrasted with others. Then, we are introduced to a potential problem with modern interpretation of literary works: modern bias. *The Tempest*, for example, could be seen as being about *colonialism* and *imperialism*; however, *conservative* critics would argue that these concepts didn't exist in Shakespeare's time because *these terms were not even in use*. By extension, the conservative view posits that Shakespeare would not have *had [these themes] in mind* when writing *The Tempest*. Instead, conservative critics claim that the play is *universal*, with *themes like estrangement and reconciliation*. From here, we want to be thinking about the direction that the passage could go—specifically, does the author agree or disagree with these conservative critics? A Label for this paragraph could be: **P1.** Conservative view: modern literary criticism has modern bias, ex: *The Tempest*

10: Question Types II

Passage	Analysis
But writers do live in specific societies and are affected by the cultures of the times and places in which they live. The establishment of colonies—the building of empires—was an issue of keen concern in England in the early 17th century. It was a matter of national prestige and also a potential source of private wealth. All the great powers of Europe were competing for the wealth of the East and West Indies. The rich hoped to add to their fortunes; the poor hoped to begin their lives anew in the New World. Richard Hakluyt's *Voyages*, a series of published accounts of European explorations in Asia, Africa, and the Americas, was one of the most successful publishing ventures of Elizabethan England. Moreover, although the words "colonialism" and "imperialism" had not been coined yet, the ideas they connote already existed, in the sense that some Europeans perceived ethical problems relating to empire-building. The Spanish priest Bartolomé de las Casas had already condemned the cruelty of the Spanish regime in Mexico and the Caribbean, and the French essayist Michel de Montaigne had already compared the "Cannibals" favorably with decadent Europeans.	The first word is the Difference keyword *but*, indicating that the author will not agree with the conservative view. To support his or her disagreement with the conservative view (and agreement with modern literary criticism), the author brings in examples. First, the author demonstrates that Europeans were indeed attentive to *the establishment of colonies—the building of empires* for a number of reasons, including *national prestige*, *wealth*, and the potential to *build…lives anew in the New World*. *Voyages* is mentioned as a popular text of the time, further supporting this first claim. Second, the author claims that although the terms *colonialism* and *imperialism* did not exist yet, the ideas they represent—*perceived ethical problems relating to empire-building*—were present in at least some Europeans' thoughts. The Spanish and French examples support this second claim. This paragraph can be Labeled as: **P2.** Author disagrees with conservatives: Europeans concerned with colonialism/imperialism even if no words for them

MCAT Critical Analysis and Reasoning Skills

Passage	Analysis
Now let us look again at *The Tempest*. Here is a play about a European family ruling a remote island by superior European technology (magic, learned from books) and the enforced labor of the native population. When another group of Europeans arrives on the island, one of them imagines an ideal commonwealth in terms derived, as scholars have long recognized, from Montaigne's essay about the native people of Brazil. The prostrate Caliban reminds Trinculo of a "dead Indian" who might be exhibited in England for crowds willing to pay to see an exotic "monster." And scholars have long recognized that the story of *The Tempest* is suggested in part by accounts of the *Sea Venture*, shipwrecked in Bermuda in 1609 on the way to the Virginia colonies.	The author returns to *The Tempest* to demonstrate his or her claims from the previous paragraph. We are given a description of the play, and the details the author chooses to focus on are reminiscent of the European concerns brought up in the last paragraph about empire-building. *Enforced labor* is reminiscent of *the cruelty of the Spanish regime* in the previous paragraph, and *Montaigne* is brought back again as an example. While the author does not explicitly say so, the implication is that these aspects of *The Tempest* represent the themes of *colonialism* and *imperialism*. A Label for this paragraph is: **P3.** Examples of colonialism/imperialism in *Tempest*
With all this in mind, are we really to believe that neither Shakespeare nor anyone who saw the play in London in 1611 was reminded of the colonial enterprise that England was then undertaking in America? Who is making the implausible claim?	This last paragraph has two rhetorical questions. The first is straightforward to answer: the author clearly believes that Shakespeare and others who watched *The Tempest* were reminded of England's *colonial enterprise...in America*. The second is a bit less obvious. The phrase *implausible claim* was used previously in paragraph 1—in that paragraph, it represents conservatives' criticism of modern literary criticism. Here, the author is saying that it is conservatives—not modern literary critics—who are *making the implausible claim*. This paragraph can be Labeled as: **P4.** Auth: Shakespeare, others recognize themes—conservatives are wrong

10: Question Types II

Here's a sample Outline and Goal for this passage:

P1. Conservative view: modern literary criticism has modern bias, ex: *The Tempest*

P2. Author disagrees with conservatives: Europeans concerned with colonialism/imperialism even if no words for them

P3. Examples of colonialism/imperialism in *Tempest*

P4. Auth: Shakespeare, others recognize themes—conservatives are wrong

Goal: To demonstrate that themes of colonialism and imperialism are present in *The Tempest* and rebut conservative critics' opposition to this view

Question	Analysis
1. It can be inferred that the author regards conservative critics of Shakespeare with:	This is an Inference question asking for the author's opinion of conservative critics. The prediction is easy to make here given the Label for the last paragraph: the author thinks that *conservatives are wrong*. Stick with this broad prediction unless the answer choices indicate that more information is needed.
A. disapproval because they ignore the influence of society on an individual.	**(A)** matches perfectly with the prediction and is the correct answer. The author's rebuttal begins in paragraph 2: *But writers do live in specific societies and are affected by the cultures… in which they live*, which further supports this answer.
B. displeasure because they defend 17th-century English colonialism.	**(B)** can be eliminated because the conservative critics' argument focuses on literary criticism, not *English colonialism* itself.
C. approval because they emphasize the universality of Shakespeare's themes.	**(C)** and **(D)** both have a positive tone, making them Opposites of the correct answer.
D. admiration because they amass facts about Shakespeare's age.	

MCAT Critical Analysis and Reasoning Skills

Question	Analysis
2. Which of the following is used in the passage to support the conservative interpretation of Shakespeare?	This is a Strengthen–Weaken (Within the Passage) question because it asks for the evidence used for a given conclusion. We know the conservative critics see Shakespeare as a timeless, *universal* playwright and not one whose themes are about only a specific time. The conservative argument in paragraph 1 hinges on the claim that the terms *colonialism* and *imperialism* did not exist yet, and therefore it would be *anachronistic to suppose that Shakespeare had them in mind*.
A. Imperialism and colonialism are anachronistic terms for the 17th century.	**(A)** fits perfectly with the prediction and is the correct answer.
B. *The Tempest* has its roots in a story regarding English ships headed for America. C. The specific society a writer lives in is essential to understanding his themes. D. European technology in the 17th century was far superior to other technology.	**(B)**, **(C)**, and **(D)** can be eliminated quickly because they do not come from paragraph 1, the only place where the conservatives' view is given any support. These answer choices come from paragraphs 2 and 3, which are not used to support the conservative view.

Question	Analysis
3. The author assumes that thematic elements:	For this Inference question of the Assumption sub-type, consider what thematic elements the author addresses. The author claims that themes in *The Tempest* center around colonialism and imperialism, even though these were not actually words present in the play or in the time period. This is a broad prediction, but we should proceed to the answer choices and use the Denial Test if necessary to check a potentially correct answer.
A. can only occur when they are able to be described by specific language. B. are present in all stories and are ubiquitous regardless of time period.	**(A)** and **(B)** both sound like the conservative critics and are inconsistent with the author's central argument. These choices should be eliminated.
C. can be present regardless of whether or not there are words to describe them.	This answer choice is very close to our prediction and can be verified using the Denial Test. First, negate the answer choice: thematic elements *cannot be present if there are not words to describe them*. If this statement were true, the author's argument would completely unravel. Therefore, **(C)** is the correct answer choice.
D. are not present in older stories where the author was living in a different time period.	**(D)** implies that older stories do not have thematic elements at all. This idea is inconsistent with the author, who argues that *colonialism* and *imperialism* are present in *The Tempest*.

Question	Analysis
4. Implicit in conservative critics' view is that:	This is an Inference question of the Implication subtype. Conservative critics think that Shakespeare is timeless and that his themes are *universal*. Further, they think that *it trivializes [Shakespeare's] genius to suppose that he had in mind the fashionable concerns of any one period*. We should predict that conservatives think Shakespeare avoided themes that go in and out of style and focused on universal themes instead.
A. Shakespeare wrote about fashionable themes.	**(A)** is an Opposite because conservative critics claim that Shakespeare avoided *fashionable themes*.
B. some thematic elements will go out of fashion.	**(B)** is a bit narrower than our prediction but still fits well. Conservative critics must believe that some thematic elements will go out of fashion—these are the themes they claim Shakespeare avoids.
C. thematic elements that go out of style come back later.	**(C)** might at first look the same as **(B)**, but it adds an additional assumption that is not supported by the passage. While themes may go in and out of style, it is a Distortion to say that thematic elements necessarily *come back later*. Some thematic elements may very well go out of style and never come back.
D. political themes are essential to good literature.	**(D)** is Out of Scope and runs counter to the conservative view; the themes conservatives identify—*estrangement and reconciliation*—are not political themes at all.

10: Question Types II

Question	Analysis
5. The fact that the terms "colonialism" and "imperialism" were not coined yet in Shakespeare's time has what effect on the author's argument?	This is another Strengthen–Weaken (Within the Passage) question. The author's idea throughout the passage is that even though the words *colonialism* and *imperialism* did not exist, these themes were still present in *The Tempest*. Therefore, the fact that these words *were not coined yet* is consistent with—but does not strengthen or weaken—the author's argument.
A. It strengthens the author's argument about Shakespeare's works. B. It weakens the author's argument about *The Tempest*. C. It strengthens the author's argument about *Voyages*.	**(A)**, **(B)**, and **(C)** can be eliminated immediately because they state that the author's argument would be *strengthen[ed]* or *weaken[ed]*.
D. It doesn't affect the author's argument.	The prediction—that the author's argument is neither strengthened nor weakened—matches **(D)**.

Practice Questions

Passage 1 (Questions 1–6)

It would be difficult to overstate the complexity of the Japanese language. The system of writing (or more properly, systems) represents a fusion of almost entirely foreign characters and a spoken language so linguistically isolated that philologists have yet to discover a precursor. Not unlike many other ancient languages, Japanese lacked any system of writing at all for much of its history. Making up for lost time, though, no fewer than three different systems of writing are now employed.

The first Japanese system of writing was not Japanese. The *kanji*, a group of logographic Chinese characters each representing a word or idea, were adopted with minimal change around the seventh century. Few languages are so geographically close yet linguistically dissimilar. As a result, the Japanese adopted a modified Chinese pronunciation for each kanji (the *on-yomi*) while retaining the native Japanese spoken word that most closely fit each kanji's meaning (the *kun-yomi*). In modern Japanese, the on-yomi is used for certain kanji and the kun-yomi for others, with compound words often involving both. Further adding to the confusion, the Chinese language contains many words in which variations in tone alone indicate drastically different meanings. The adaptation of these words to Japanese pronunciation led to a number of homophones that has, without hyperbole, been called "embarrassing" and "alarming" by scholars of the language.

The *hiragana* syllabary was developed in the eighth century by court women, who were not permitted to study kanji because they were deemed unfit to master its complexities. In response, they developed a simplified, flowing form of the kanji that represented all the sounds in spoken Japanese. Hiragana is phonetic rather than logographic and is therefore far more accessible to a foreign learner than the kanji. Because Japanese is an open language, most consonants cannot be expressed by themselves. Hiragana is therefore not strictly an alphabet. *Katakana* came about around the same time as hiragana, also as an attempt to simplify the kanji. The sparse, angular characters correspond fairly closely to the hiragana and, as befitting their origin among Buddhist monks, have a look generally considered more masculine than hiragana, which was originally called *onnade*, or "women's hand." The katakana have essentially become the print counterparts of the "cursive" hiragana.

With so many systems jostling for position, each used more or less independently of the other, it would not be unreasonable to anticipate that a national movement towards systematization of the language would settle on a single one. A national movement was in fact started after World War II: a radical idea encouraging the use of all three systems together. A glance at any Tokyo newspaper will reveal kanji used to represent most standard actions and ideas, hiragana to indicate grammatical inflections and tenses, and katakana to represent adopted foreign and technical words, as well as to indicate emphasis. The use of the three systems has become sufficiently standardized in this way that deviations often lend a piece of writing strong connotations. A piece written entirely in katakana, for example, may be disconcerting to a modern reader and may have a vaguely pre-World War II military air to it, much as a piece written all in capital letters with telegraph punctuation might in English. While such a complex system has made the language's learning curve high for native speakers and foreigners alike, it has also contributed to a stunning richness of expression such that any list of world's great works of art a hundred years from now will have to be written partially in kanji, hiragana, and katakana.

1. The author's explanation for the origin of Japanese homophones relies on which of the following assumptions?

    A. The large number of homophones in the language is due to the closeness of Japanese and Chinese pronunciation.
    B. The presence of homophones in a language can be considered embarrassing.
    C. Spoken Japanese does not rely on tone to the extent that Chinese does.
    D. Homophones are dependent on variations in tone.

2. The author's primary purpose in the passage is to:

    A. argue that the Japanese language is overly complex.
    B. describe the origins of the Japanese language's complexity.
    C. propose a simplification in how Japanese is written.
    D. trace the origins of logographic writing systems.

3. According to the passage, which of the following pieces of Japanese literature would NOT likely be written entirely in katakana?

    A. a modern Japanese novel
    B. a list of adopted foreign words
    C. a ninth-century Buddhist text
    D. an early 20th-century general's log

4. Based on the author's description, open languages generally contain:

    A. borrowed systems of writing and speaking from many different sources.
    B. intrinsic acceptance of change and reform.
    C. syllables that end in vowels.
    D. few consonant sounds.

5. As can be inferred from the passage, the group of Buddhist monks who developed katakana:

    A. was predominantly or entirely male.
    B. used hiragana as a model.
    C. was considered unfit to master the complexities of kanji.
    D. was closely involved with the military of the time.

6. Which of the following is a claim the author makes without providing evidence?

    A. The first Japanese system of writing was not Japanese.
    B. The use of the three systems has become sufficiently standardized in this way that deviations often lend a piece of writing strong connotations.
    C. Japanese lacked any system of writing at all for much of its history.
    D. The katakana have essentially become the print counterparts of the "cursive" hiragana.

Passage 2 (Questions 7–12)

The palette of sights and sounds that reach the conscious mind are not neutral perceptions that people then evaluate: they come with a value already tacked onto them by the brain's processing mechanisms. Tests show that these evaluations are immediate and unconscious and applied even to things people have never encountered before, like nonsense words: "juvalamu" is intensely pleasing and "bargulum" moderately so, but "chakaka" is loathed by English speakers. These conclusions come from psychologists who have developed a test for measuring the likes and dislikes created in the moment of perceiving a word, sound, or picture. The findings, if confirmed, have possibly unsettling implications for people's ability to think and behave objectively. This is all part of preconscious processing, the mind's perception and organization of information that goes on before it reaches awareness—these judgments are lightning fast in the first moment of contact between the world and the mind.

Some scientists disagree with the claim that virtually every perception carries with it an automatic judgment, though they, too, find that such evaluations are made in many circumstances. This cohort posits a narrower scope of stimuli that elicit the response. That is, these scientists believe that people don't have automatic attitudes for everything, but rather for areas of interest.

In responding to a stimulus, a signal most likely travels first to the verbal cortex, then through white matter tracts to the amygdala, where the effect is added, and then back to the occipital lobe through the same or similar pathways. The circuitry involved can do all of this in a matter of a hundred milliseconds or so, long before the individual experiences any conscious awareness of the word. This creates an initial predisposition that gets things off on a positive or negative footing. These reactions have the power to largely determine the course of a social interaction by defining the psychological reality of the situation from the start.

Although perhaps counterintuitive, the "quick and dirty" judgment tends to be more predictive of how people actually behave than is their conscious reflection on the topic. This may represent a new, more subtle tool for research on people's attitudes, allowing scientists to assess what people feel without their having any idea of what exactly is being tested. One could detect socially sensitive attitudes people are reluctant to admit, like racial and ethnic biases, because these automatic judgments occur outside of a person's awareness, as part of an initial perception. They are trusted in the same way senses are trusted, not realizing that seemingly neutral first perceptions are already biased.

Conclusions from both camps are based on a method that allows them to detect subtle evaluations made within the first 250 milliseconds—a quarter of a second—of the perception of words. The measurement of liking can be made outside the person's awareness because if the first word is presented in less than a quarter of a second, the reaction to it never registers in consciousness, though it can still be read.

7. The author's description of reactions to words like *juvalamu*, *bargulum*, and *chakaka*, relies on which of the following assumptions?
   A. These words can be pronounced in less than 250 milliseconds.
   B. The meaning of a word is not necessary for an emotional response to it.
   C. Familiarity with a word can cloud judgment of one's reaction to the word.
   D. An individual could repeat back the words after hearing them.

8. Which of the following, if true, would serve to most strengthen the argument of an opponent to the author?
   A. Many of our actions are influenced by perceptions unknown to our consciousness.
   B. In Swahili, "juvalamu" and "chakaka" mean "enjoyable" and "severe pain," respectively.
   C. People's actions are most regulated by conscious thought patterns rather than unknown feelings.
   D. Humans perceive their surroundings subjectively.

9. The view of the cohort of scientists mentioned in paragraph 2 is most challenged in the passage by:
   A. the description of the neural circuits involved in responding to a stimulus.
   B. the claim that the "quick and dirty judgment" is more predictive of behavior than conscious reflection.
   C. the referenced study of reactions to nonsense words in English speakers.
   D. the hypothesis that measurements can be made outside of a person's awareness.

10. Based on the passage, information retrieved from these types of perception experiments could best be used by psychologists to:
    A. help patients with language barriers.
    B. map out the pathological thought patterns in a murderer's mind.
    C. identify hidden attitudes that cause two individuals to repeatedly clash on various issues.
    D. determine why one sibling has math skills while the other excels in literature analysis.

11. Based on information in the passage, in the author's view, which of the following statements is NOT true?
    A. Information regarding external stimuli is processed so quickly that it does not become part of our conscious awareness.
    B. Automatic judgments occur on stimuli with which there is great familiarity.
    C. Automatic judgments have little effect on a person's mood.
    D. Ethnic biases may be influenced by attitudes of which we are unaware.

12. If given the chance to expand on the points put forth in the passage, the author would most likely argue:
    A. to use this type of experimentation to map the pathway through which neurological signals travel.
    B. that the evidence presented in the passage is inconclusive and directs psychologists in no specific direction.
    C. that automatic judgments have little or no effect on important behavior patterns.
    D. to continue with further experimentation, seeking to identify the roots of problems found in human relationships.

Explanations to Practice Questions

Passage 1 (Questions 1–6)

Sample Passage Outline

P1. Japanese language and writing complex: at least 3 systems

P2. Chinese (logographic) introduced = kanji; on-yomi *vs.* kun-yomi

P3. Syllabic/phonetic: hiragana (feminine, script) *vs.* katakana (masculine, print)

P4. Post-WWII: All three combined and systematized, different uses for each

Goal: To examine the origins and complexity of three Japanese writing systems

1. C

For this Inference question of the Assumption subtype, start with where *homophones* are mentioned. Homophones appear in paragraph 2, where the author says that the adaptation of Chinese words with variations in tone to Japanese resulted in lots of homophones. Break this argument down to its constituent parts: the conclusion is that Japanese has many homophones, and the evidence is that the words were adapted from a language where tone changes meaning. To find the inherent assumption, we need to determine what additional detail must be true in order to come to that conclusion. To bridge the two ideas together, we can predict that Japanese doesn't have as many variations in tone as Chinese does. **(C)** fits this prediction. While **(B)** may be a true statement, it is not related to *the origin of Japanese homophones* and is a Faulty Use of Detail answer choice. **(D)** is an Opposite; it is the loss of *variations in tone* as words moved from Japanese to Chinese that led to the formation of homophones. As for **(A)**, the author specifically stated that *Few languages are so geographically close yet linguistically dissimilar*, so we know that pronunciations must be very different.

2. B

This is a Main Idea question, so predict using the author's overall Goal: *to examine the origins and complexity of three Japanese writing systems*. Only **(B)** involves both the *origins* and *complexity* of the language. Notice that we can use a vertical Scan to eliminate **(A)** and **(C)** because the author is neutral and does not make any strong *argu[ments]* or *propos[als]*. As for **(D)**, the author discussed the origins of Japanese only—not multiple *logographic writing systems*—and, even then, this answer choice is too narrow as it addresses the author's purpose only in paragraph 2.

3. A

This is an Apply question asking for an Example of a text that would *NOT* likely be written entirely in *katakana*. Where does the author discuss the uses of writing in katakana? Looking at the Outline, we find that it is introduced in paragraph 3 and that the modern approach of using all three systems together—including katakana—is described in paragraph 4. There, the author writes that katakana is used to *represent adopted foreign and technical words, as well as to indicate emphasis* and that a piece written entirely in katakana *may be disconcerting to a modern reader* and would have a *pre-World War II military air* to it. Based on this information, **(B)** and **(D)** can immediately be eliminated. **(C)** can also be eliminated based on the description of the origins of katakana in paragraph 3: both hiragana and katakana appear around the *eighth century*, and katakana specifically *origin[ated] among Buddhist monks*. The answer must therefore be **(A)**, which makes sense: a modern piece of literature would be expected to combine all three writing systems.

4. C

For this Inference question of the Implication subtype, start with where *open languages* are mentioned. Paragraph 3 states that *because Japanese is an open language, most consonants cannot be expressed by themselves*. It also points out that hiragana (and, by extension, katakana) is *not strictly an alphabet*, but rather a *syllabary*. Taking these pieces of information together, we can determine that open languages must express consonants together with vowels and that the language is built on these consonant–vowel combinations (syllables). **(C)** reflects this idea, highlighting the syllabic nature of the language. While the author does not specifically state that vowels end syllables in Japanese, the two Japanese terms given in the paragraph—hiragana and katakana—both demonstrate this pattern. **(D)** is a Distortion because although the author does say that consonants are not often used by themselves, there is no mention that they are few in number overall. While **(A)** describes the Japanese language, it does not reflect the author's use of the more general term *open languages*. Finally, **(B)** is a literal use of the word *open* and does not fit the context described by the author.

5. A

The word *inferred* shows that this is an Inference question. The Buddhist monks who developed katakana are highlighted in paragraph 3. Let's review the main points: katakana was created by *Buddhist monks* and looks *more masculine than hiragana*. The author also notes that hiragana was developed by women and was known as *women's hand*. Given the contrast between katakana and hiragana on the basis of gender, we can infer that katakana looks masculine because it was developed by men. **(A)** must therefore accurately describe this group of Buddhist monks. While **(C)** might look tempting, we know only that the women who developed hiragana were *considered unfit to master the complexities of kanji*. The author never stated anything similar about the monks who developed katakana, so we cannot make that inference. In the last paragraph, the author points out that katakana now has a *vague…military air* to it for the modern reader, but that does not mean that the monks who created katakana were *closely involved with the military of the time*.

6. C

This is a Strengthen–Weaken question asking for a claim in the passage that lacks evidence. Note that all four answer choices are sentences, taken verbatim, from the passage. For this question, we will have to address each answer choice, as we look for evidence that supports the claim. The *first Japanese system of writing*, **(A)**, was addressed in paragraph 2. Immediately following this sentence, the description of *kanji* is given—which is that *first…system of writing*. Therefore, this answer choice can be eliminated. The use of *three systems* and the effects of *deviations* from the standardized approach are detailed in paragraph 4; the subsequent *piece written entirely in katakana* is evidence to support this claim, eliminating **(B)**. **(C)** is mentioned at the end of paragraph 1, but that's all the information we get about the Japanese language before writing systems were developed. This answer is therefore correct. Finally, **(D)** is supported by the sentence that immediately precedes it, which describes the *sparse, angular characters* of katakana as counterparts to the hiragana.

MCAT Critical Analysis and Reasoning Skills

Passage 2 (Questions 7–12)

Sample Passage Outline

P1. Value assigned to perceptions before reaching awareness (preconscious processing)

P2. Some scientists think scope is narrower: areas of interest

P3. Brain pathway of preconscious processing

P4. Applications of the theory, ex: socially sensitive attitudes and biases

P5. Methods of measuring

Goal: To analyze the role of preconscious processing on attitudes toward words and ideas

7. B

This is an Inference question of the Assumption subtype focusing on the author's example in the first paragraph. In this *nonsense words* example, words devoid of any importance in the English language appear to trigger an emotional response. This might be surprising—usually we would think that any emotional response to language would be based on the meaning of the word. However, if individuals still have responses to these nonsense words, knowing the definition of the word must not be required. This assumption matches closely with **(B)**, making it the correct answer. **(A)** is a Faulty Use of Detail; while the last paragraph states that a word would have to be given in less than 250 milliseconds to avoid conscious perception, the example in the first paragraph does not require that the word never reaches conscious perception. **(C)** is a Distortion. While this answer choice may very well be a true statement, the example described in the first paragraph does not address familiar words at all—only nonsense words; thus, we cannot infer anything about familiar words for this example. Finally, **(D)** is Out of Scope; whether or not the individual can repeat back the words has no clear bearing on the individual's feelings about the word.

8. C

The words *if true* and *strengthen* indicate that this is a Strengthen–Weaken (Beyond the Passage) question, which is similar to the Strengthen–Weaken (Within the Passage) question type discussed in this chapter, except it brings in new information. The question asks to *strengthen the argument of an opponent to the author*, which would logically be the same as weakening the author's conclusion. The author's main conclusion in the passage is that many impressions are formed preconsciously, so any answer choice that goes against this premise would be correct. **(C)** says exactly the opposite of what the author argues, claiming that it is *conscious thought*—not *unknown feelings*—that dictates behavior. **(A)** and **(D)** both fit cleanly with the author's argument and therefore would not weaken his or her conclusion. **(B)** is Out of Scope as the author does not address the responses to these nonsense words in any populations besides English speakers.

9. C

This Strengthen–Weaken (Within the Passage) question requires a bit of decoding. The *cohort of scientists mentioned in paragraph 2* believes that automatic attitudes do not occur in response to *everything, but rather…areas of interest*. To challenge this claim, we would need some evidence of an automatic attitude formed in response to something unlikely to be a person's area of interest. We find a match to this prediction in **(C)**; nonsense words are unlikely to be an area of interest for most people, and yet an automatic response was still generated. The other answer choices all bring in other components of the passage, but the *neural circuits*, *predict[ability] of behavior*, and methods for *measurements* all have nothing to do with the *areas of interest* claim.

10. C

This is an Apply question asking for an Example of the possible uses for the *information retrieved from the…perception experiments* described in the passage. Possible uses of this information are explored in paragraph 4, which is

Labeled: *Applications of the theory, ex: socially sensitive attitudes and biases*. Our prediction, then, is an answer that identifies *attitudes and biases*; this matches best with **(C)**. While the example of racial and ethnic biases is mentioned in the passage, other *attitudes people are reluctant to admit* are also considered. The other answer choices are all Out of Scope, as they bring in *language barriers, pathological thought patterns*, and differences in cognitive abilities between *siblings*—none of which is even hinted at in the passage.

11. C

Here we've got another Inference question, but this one is a Scattered Inference question due to the *NOT* in the question stem. Keep the author's main conclusion in mind before looking at the answer choices—that incoming information has value placed on it before we're even aware of it. Three answers will agree with the author's argument, and one should work against it. In this case, **(C)** is unusually easy to spot: it contradicts the basic conclusion of the experiment the author cites in paragraph 1. In this experiment, automatic judgments placed on nonsense words appeared to generate an emotional response. **(A)** and **(D)** both focus on the author's point that we are unaware of the processing that is happening. **(B)** might also be acceptable to the author; just because *automatic judgments occur on stimuli* with which one is unfamiliar doesn't mean that these judgments won't occur on stimuli that are familiar.

12. D

For this Strengthen–Weaken (Within the Passage) question in which we must continue an argument already present in the passage, consider the word *argue*: whatever the author argues will have to match the ideas he or she already presented in the passage. First, knock out any answer choices that weaken the author's stance; this allows us to eliminate **(B)** and **(C)**, as they would ruin the author's central thesis. Then, **(D)** is a logical extension of what the author describes for the majority of the passage, so it's the best fit. While the author does mention the *pathway* in **(A)**, this idea is a small point in the passage; the primary focus is on psychology, not neuroanatomy.

Question Types III: *Reasoning Beyond the Text* Questions

11: Question Types III

In This Chapter

11.1 Apply Questions — 266
    Sample Question Stems — 267
    Strategy — 268
    Worked Example—
        A Psychology Passage — 268

11.2 Strengthen–Weaken (Beyond the Passage) Questions — 273
    Sample Question Stems — 273
    Strategy — 273
    Worked Example—
        An Arts Passage — 274

11.3 Other *Reasoning Beyond the Text* Questions — 277
    Probable Hypothesis — 277
    Alternative Explanation — 278
    Passage Alteration — 278

Concept and Strategy Summary — 280

Worked Example — 283

Introduction

> **LEARNING GOALS**
>
> After Chapter 11, you will be able to:
>
> - Identify Apply, Strengthen–Weaken, and Other *Reasoning Beyond the Text* questions
> - Solve *Reasoning Beyond the Text* questions with strategies specific to each question type
> - Differentiate between Probable Hypothesis, Alternative Explanation, and Passage Alteration questions

If patients came with instruction manuals, medical school might not be necessary. Many of the clinical tasks you will perform as a physician require applying the vast quantities of knowledge you will master over your years of medical school to situations that never quite match the ones you read about in class. When real-life patients are conscious and able to describe their symptoms, the language they use rarely coincides with the sophisticated terminology employed in textbooks and journals. Consequently, much of medical training involves actual patient interactions, transcending the classroom to include lessons that can be taught only at the bedside. Medical schools are interested in students who know how to go beyond what they've read, which is why the *Critical Analysis and Reasoning Skills* (CARS) section makes this a major focus. Although it may not be the intensive care unit, the MCAT is like the ICU in that both test whether you understand what you think you've learned well enough to apply it with both speed and precision.

MCAT Critical Analysis and Reasoning Skills

In this chapter, we'll examine the Apply and Strengthen–Weaken (Beyond the Passage) question types. As in the previous question types chapters, we'll look at some common question stems, specific strategies, and a few worked examples for each question type. We conclude with a brief discussion of rarer kinds of *Reasoning Beyond the Text* questions.

The AAMC reports that 40 percent of the questions in the CARS section should be classified as *Reasoning Beyond the Text*, and it further divides this categorization into questions that require you to apply or extrapolate ideas from the passage to a new context (Apply questions) and those that require you to evaluate the effect new information would have if incorporated into the passage. (Because the effect is usually a strengthening or weakening, we call these Strengthen–Weaken (Beyond the Passage) questions.) In other words, the fundamental difference is one of direction: Apply questions go from passage to new situation, while Strengthen–Weaken (Beyond) questions go from new situation to passage. While the AAMC seems to suggest that the *Reasoning Beyond the Text* category is split evenly between these, according to our intensive study of released AAMC materials, Apply questions appear to be a bit more common than Strengthen–Weaken (Beyond) questions. So, out of the approximately 21 *Reasoning Beyond the Text* questions that you'll see in a CARS section, typically about 11 will be Apply questions, and only about 8 or 9 will be Strengthen–Weaken (Beyond) questions, with perhaps 1 that doesn't neatly fall into either type.

Note: The Question Types, as well as the Kaplan Method for CARS Passages, Kaplan Method for CARS Questions, and Wrong Answer Pathologies are included as tear-out sheets in the back of this book.

11.1 Apply Questions

Questions that fall into the broad category of *Reasoning Beyond the Text* are easy to identify because they always involve novel information (in the question stem, set of answer choices, or both) that is not stated or even suggested by the passage and that may not even seem to be related at first. In one typical pattern, words like *Suppose*, *Assume*, and *Imagine* precede an elaborate scenario that fills several lines of text, followed up with a question that connects this new content to the author or passage. There are two basic ways in which this connection may be asked about, varying with respect to the direction in the relationship: passage → new situation or new situation → passage. The first relationship makes up the Apply question type, while the second constitutes Strengthen–Weaken (Beyond) questions.

Apply questions are the most common of the eight major types discussed in these three chapters. They take the text as a starting point and ask you to extrapolate to a new context. While the questions posed can vary considerably, there are three

> **Bridge**
>
> Apply and Strengthen–Weaken (Beyond the Text) are examples of deductive reasoning at its finest. In both cases, a new situation is provided. In Apply questions, we focus on how the passage relates to the new information (through a Response, Outcome, or Example). In Strengthen–Weaken (Beyond) questions, we focus on how the new information impacts the passage. Deductive reasoning and other problem-solving techniques are discussed in Chapter 4 of *MCAT Behavioral Sciences Review*.

frequently occurring tasks, each of which constitutes roughly one-third of the Apply question pool: we call them *Response*, *Outcome*, and *Example*, reflecting some of the words they commonly contain.

SAMPLE QUESTION STEMS

- Consider the following: [new info]. The author would most likely respond to this by claiming:
- With which of the following claims would [the author or an alternative viewpoint from the passage] be LEAST likely to agree?
- Suppose that [details of new scenario]. Based on the passage, what would the author most probably advise in such a scenario?
- Imagine [new info]. Which of the following, according to information presented in the passage, is the most reasonable outcome?
- If the passage's author is correct, the most likely consequence of [new situation] would be:
- Assume that [new info]. One could reasonably expect, on the basis of the passage, that:
- Which of the following best exemplifies the author's notion of [quotation from the passage]?
- Which of the following phenomena would the author most likely characterize as a [concept]?
- [New info]. The author would most likely classify this as:

At least a third of Apply question stems are similar to the first three samples above, concerning how the author (or, less frequently, some other individual or a proponent of a theory discussed in the passage) would respond to a particular situation. Besides asking for the likely *response* or *reply*, these questions may simply ask for a claim that the author would be *most likely to agree with* or the statement *least consistent with* one of the views discussed.

Other Apply question stems are like the next three samples above, investigating the most probable *outcome*, *result*, *expectation*, or *consequence* in a situation that is in some way analogous to that discussed in the passage. In other words, these questions give you a cause and ask about what effect is likely if the passage is believed.

Finally, most of the remaining third of Apply question stems resemble are like the final three cases, asking for *examples* or *instances* of ideas discussed in the passage. More often, the concept will be given, and you'll have to find an item from a specified context (or from the "real world") that would provide the best example of that term as the author uses it. Rarer and more difficult are questions like the last one, which start with the outside case and ask you how the author would categorize it: *What is this an example of?*

MCAT Expertise

According to our research of released AAMC material, Apply questions make up about 21 percent of the CARS section (about 11 questions).

MCAT Expertise

Whenever a question stem is particularly lengthy, it is almost certainly going to involve *Reasoning Beyond the Text*, but it could just as easily be Apply as Strengthen–Weaken (Beyond). To save time on Test Day, whenever you see many lines of new information, perhaps preceded by clues like *Suppose*, *Imagine*, or *Assume*, take a moment to jump ahead to the very last line before the question mark or colon at the end of the question stem to determine which of the two types of questions this is and precisely what it is that you'll be looking for. After you have a better idea of your task, Plan to read through the new information carefully, always watching out for analogies to and similarities with the passage text.

MCAT Critical Analysis and Reasoning Skills

STRATEGY

If new information is provided in the question stem, you may find it advantageous to jump to the colon or question mark at the end to see what the question is really asking. Once you've picked up on the key language to Assess your task, read any new information in the question stem closely for hints that connect it to the passage. Then go back and reread the relevant portions to make your prediction. When Planning, keep in mind the type of Apply involved.

Response

If the stem asks for how the author would Respond or a claim the author would be likely to endorse, your task will be to get inside the author's head. The correct answer to a Response question should be consistent with the author's beliefs, which are typically reflected in the passage through the use of Author keywords, originally discussed in Chapter 3 of *MCAT CARS Review*. If you find yourself with an unhelpful prediction, eliminate any answers that would be logically inconsistent with the author's assertions. If a viewpoint other than the author's is asked about, follow a similar strategy by putting yourself in the mindset of the alternative perspective mentioned in the stem.

Outcome

With questions that ask about the probable Outcomes of scenarios, you'll want to pay more attention to language in the passage that deals with cause and effect, much of which will be Logic keywords. Identify any causes in the passage that are analogous to what is presented in the question stem and use the effects for which they are said to be responsible as the basis for your prediction.

Example

In those cases when you are called upon to identify Examples, find the relevant text from the passage, especially any text that provides definitions, explanations, or the author's own examples of the concept in question. Take note of necessary conditions (which must occur in all instances of the concept) and sufficient conditions (which are enough on their own to make an instance qualify as that concept), both originally discussed in Chapter 6 of *MCAT CARS Review*. Sufficient conditions allow for quicker matches, but necessary conditions can be used to rule out answers that lack them in a process of elimination.

WORKED EXAMPLE—A PSYCHOLOGY PASSAGE

There is no shortage of evidence for the existence of systemic biases in ordinary human reasoning. For instance, Kahneman and Tversky in their groundbreaking 1974 work proposed the existence of a heuristic—an error-prone shortcut in reasoning—known as "anchoring." In one of their most notable

11: Question Types III

experiments, participants were exposed to the spin of a roulette wheel (specially rigged to land randomly on one of only two possible results) before being asked to guess what percentage of United Nations member states were African. The half of the sample who had the roulette wheel stop at 65 guessed, on average, that 45% of the UN was African, while those with a result of 10 guessed only 25%, demonstrating that prior presentation of a random number otherwise unconnected to a quantitative judgment can still influence that judgment.

The anchoring effect has been observed on repeated other occasions, such as in Dan Ariely's experiment that used digits in Social Security numbers as an anchor for bids at an auction, and in the 1996 study by Wilson *et al.* that showed even awareness of the existence of anchoring bias is insufficient to mitigate its effects. The advertising industry has long been aware of this bias, the rationale for its frequent practice of featuring an "original" price before showing a "sale" price that is invariably reduced. Of course, anchoring is hardly alone among the defective tendencies in human reasoning; other systemic biases have also been experimentally identified, including loss aversion, the availability heuristic, and optimism bias.

Example:

1. Suppose a consumer who is looking for an inexpensive replacement for her outmoded refrigerator is drawn by a local retailer's ads for a discount sale promising savings of 50 percent or greater on all appliances. The author would probably warn the consumer that:

   A. sales are scams designed to exploit the consuming public.
   B. the pre-markdown prices are most likely set artificially high.
   C. heavily discounted merchandise is likely damaged or stolen.
   D. making a rational decision about what to buy is impossible.

Solution: With the lengthy question stem, it's worth your while to jump to the part immediately before the colon to Assess what kind of *Reasoning Beyond the Text* will be required. In this case, you're tasked with determining what the author would warn, a kind of Response, which means you want to be thinking about views that the author holds. The ads for the sale should draw you to paragraph 2, in which the author states: *The advertising industry has long been aware of this bias, the rationale for its frequent practice of featuring an "original" price before showing a "sale" price that is invariably reduced.* Although the author is not explicit, the use of quotation marks here is a case of "scare quotes," suggesting that the so-called "original" price is just there to make the "sale" price seem lower. Thus, it's reasonable to infer that the author would warn the consumer about the anchoring effect intended with presale prices.

The closest match to this prediction is **(B)**, which actually comes quite close. **(A)** can be ruled out as a Distortion because the language is just too strong. Any advice or admonitions that an author would provide should be consistent with what the author says in the passage, but they should also share a similar tone. The author here is not so condemning of this advertising practice as to endorse a suggestion like **(A)**. **(D)** is also a Distortion because of the word *impossible*. Finally, **(C)** is Out of Scope. The author suggests that discounts are offered to manipulate buyers into believing they have found a better deal, not to trick them into buying products that turn out to be faulty.

Example:

2. Imagine that a psychologist specialized in the study of systemic reasoning biases. On the basis of the information presented, this psychologist could most reasonably be expected to:
   A. have a higher likelihood of misjudging numerical quantities when not given an anchor.
   B. make significantly fewer mistakes in reasoning than those ignorant of anchoring bias.
   C. be equally as susceptible to errors resulting from the anchoring effect as anyone else.
   D. avoid entirely the logical fallacies that ordinary human beings commit systematically.

Solution: The question stem is relatively short, so there's no need to jump to the end before reading the part after *Imagine*. Because we're searching for something *this psychologist could most reasonably be expected to* do, you can Assess the task as finding an Outcome. Does the passage suggest anything about what happens to psychologists with greater knowledge of these biases in reasoning? There is no explicit reference, and the question stem is not heavy on other clues to facilitate much of a prediction. With a question like this, Plan to start examining the answer choices to look for additional clues about how the passage can help you.

The very first possibility suggests a consequence that would occur without the presentation of an anchor. However, the passage only concerns what happens when an anchor is present; it tells us nothing about how accurate people's judgments of quantities are without the anchoring effect. We therefore have zero basis for deciding what effect having no anchor has. Eliminate **(A)** as Out of Scope.

(B) suggests that such a psychologist would make many fewer mistakes because of her knowledge of the anchoring effect. Is there anything in the passage that would warrant such a conclusion? In fact, you discover quite the contrary. In the second paragraph, the author states that Wilson and his colleagues *showed even awareness of the existence of anchoring bias is insufficient to mitigate its effects*. Not only does this rule out **(B)** as an Opposite, but it also gives you an idea of what would make for a correct answer—one that says this psychologist is just as likely as anybody else to fall into these errors due to anchoring.

This revision to your Plan pays off when you read **(C)**, which is almost an exact match for the new prediction. On Test Day, you would select this choice and move on to the next question without paying much attention to **(D)**—which could be ruled out for being a Distortion or an Opposite, constituting an even more extreme version of **(B)**.

Example:

3. Which of the following would the author be LEAST likely to consider a case of anchoring bias?
   A. An unusually high opening bid at an annual charity auction leads to a sizable increase over previous years in total proceeds collected.
   B. The sequel to a popular film is deemed a failure because it could not quite beat the record-smashing box office receipts of the original.
   C. A shipping website receives reports of greater levels of customer satisfaction after starting deliberately to overestimate delivery times.
   D. Poor initial sales figures for a new video game console motivate its manufacturer to reduce significantly the system's suggested retail price.

Solution: Despite asking for the *LEAST likely* case, this is still an Example question. While it's hard to know what to predict based on the limited information in the question stem, you can still set expectations about the correct answer by Planning to investigate the concept mentioned. The anchoring effect is the primary subject of this short passage, but the author only provides examples of the phenomenon without giving it an explicit definition. It's difficult to say precisely what would be sufficient to constitute anchoring bias, but we can isolate some necessary conditions. In each of the passage examples, a baseline numerical expectation or "anchor" is set (either at random, as in the case of the experiments mentioned, or deliberately

MCAT Expertise

Although we generally recommend as a first Plan of attack that you make a prediction before looking at any of the answer options, with *Reasoning Beyond the Text* questions, which contain new elements in the answers, sometimes looking at the first one or two answer choices can give you a better idea of the form that the correct answer will take. If you find that the options diverge significantly from what you expected, it's advisable to go back and modify your original expectations before moving on to any remaining choices. As explained in Chapter 8 of *MCAT CARS Review*, matching to the correct answer is generally less time-consuming than crossing out all three incorrect choices, so revising your prediction is usually a more optimal strategy than process of elimination.

high in order to manipulate purchasers), which then skews the judgments people make about quantities. Any answer choice that satisfies these prerequisites should be eliminated because you are asked to find the *LEAST likely* example.

(A) is clearly a case of the anchoring effect; the passage even made mention of an experiment that used bids at an auction as the dependent variable. You can reason that the lofty opening bid must have caused other participants to heighten their appraisals of the items for sale, which in turn led to larger final sale prices and increased total proceeds.

Even though **(B)** is unlike anything found directly in the passage, it nevertheless seems to follow the model we anticipated for wrong answers. In this case, the original film is serving as the anchor that biases judgments of its sequel. That the new movie *could not quite beat the record-smashing box office receipts of the original* suggests that it still generated a lot of revenue, meaning that the assessment of failure was probably in error. Because the anchoring effect is said to be a systemic bias, this implication of error should quell any remaining doubts about eliminating **(B)**.

Turning to **(C)**, we find another case that departs considerably from the passage. Here, the anchor would be the estimated time of arrival for a particular shipment. If the site deliberately overestimates shipping times, that means its customers will consistently have to wait less time than they are told to expect, which will lead many to think that they are receiving excellent service. This situation is precisely analogous to the example of the original price marked down for sale, and hence it should also be eliminated.

This entails that the correct answer is **(D)**. While similar to one of the examples discussed on a superficial level, it does not fit the model of anchoring bias that we articulated. The monetary value of the sales (which would undoubtedly be orders of magnitude *higher* than the original price of one individual console) does not *bias* the manufacturer to lower the cost per system. Rather, this is a case of a rational response to an economic problem: when demand is too low, reduce the price.

Now, to be clear, this reduction in price could *lead to* a case of anchoring bias if, say, consumers started to purchase the console in greater quantities, believing it now to be a better deal. But the answer choice does not focus on that effect nor on any of the effects of cutting the cost, rather only mentioning its cause. And so, **(D)** is indubitably the one *LEAST likely* to count as anchoring bias for the author.

11: Question Types III

11.2 Strengthen–Weaken (Beyond the Passage) Questions

Like the similarly named category detailed in the preceding chapter, **Strengthen–Weaken (Beyond the Passage) questions** concern evidence–conclusion relationships. However, unlike the other type of Strengthen–Weaken questions, at least one of the claims involved will not be from the passage but will be unique to the question stem or answer choices. Strengthen–Weaken (Beyond) questions are also distinct because they treat the passage as flexible, subject to modification by outside forces.

SAMPLE QUESTION STEMS

- Suppose [new info]. This new information:
- Which of the following statements, if true, would most bolster the author's argument about [topic]?
- Assume that [new info]. This assumption weakens the author's claim that:
- [New info] would most strongly support the view of:
- Recent research on [topic] suggests [new info]. Which of the following assertions from the passage is most logically consistent with these results?
- Some theorists have argued that [new info]. Based on the discussion in the passage, which of the following would present the greatest CHALLENGE to their argument?
- [New info]. In conjunction with information presented in the passage, it would be most reasonable to conclude that:
- Imagine that [new info]. What impact would this have on the arguments made in the passage?
- Which of the following study findings would most seriously undermine the author's thesis?

If the question includes new information and asks about logical relationships like support, challenge, and consistency, you can safely Assess that it's a Strengthen–Weaken (Beyond) question. However, sometimes the new information is hidden in the answer choices, so you may have to watch out for clues that suggest the correct answer will come from outside the passage, such as words like *would* and *could*.

STRATEGY

As with Strengthen–Weaken (Within the Passage) questions, your primary task will be to identify the three relevant parts: the conclusion, the evidence or refutation, and the nature of the connection (strengthen, weaken, or some unspecified relevance). Your first step should be to determine which component (or, rarely, which two components) you'll be seeking. If the stem is long, don't bother to read all the

> **MCAT Expertise**
>
> According to our research of released AAMC material, Strengthen–Weaken (Beyond the Passage) questions make up about 16 percent of the CARS section (about eight or nine questions).

MCAT Critical Analysis and Reasoning Skills

Bridge

Strengthen–Weaken (Beyond the Passage) questions are extremely similar to Strengthen–Weaken (Within the Passage) questions, except that the former bring in new information while the latter ask about arguments wholly contained in the passage. Accordingly, the strategic approaches to these two types of questions are very similar. Make sure to review Strengthen–Weaken (Within) questions, discussed in Chapter 10 of *MCAT CARS Review*, in tandem with this discussion of Strengthen–Weaken (Beyond) questions.

new details the first time through. Instead, jump to what immediately precedes the question mark or colon in order to figure out where the question stem is going. With this Plan in mind, turn next to what you are given, reading the entire question stem closely now and keeping an eye out for any hints of analogy. So, for instance, if a novel experimental finding described in the question stem reminds you of a study in the passage that was used to support the author's thesis, then chances are that the correct answer will indicate that this *strengthens* the thesis, or a similar idea.

From this point, the same strategy considerations apply as did to Strengthen–Weaken (Within) questions, with Logic keywords from the passage again playing a major role. The major differences will be that correct answers to Strengthen–Weaken (Beyond) questions are seldom exact matches to your predictions and are far more likely to be only incidentally related to the text.

WORKED EXAMPLE—AN ARTS PASSAGE

One of the first examples of the ascendance of abstraction in 20th-century art is the Dada movement, which Lowenthal dubbed "the groundwork to abstract art and sound poetry, a starting point for performance art, a prelude to postmodernism, an influence on pop art...and the movement that laid the foundation for surrealism." Dadaism was ultimately premised on a philosophical rejection of the dominant culture, which is to say the dominating culture of colonialist Europe. Not content with the violent exploitation of other peoples, Europe's ruling factions once again turned inward, reigniting provincial disputes into the conflagration that came to be known by the Eurocentric epithet "World War I"—the European subcontinent apparently being the only part of the world that mattered.

The absurd destructiveness of the Great War was a natural prelude to the creative absurdity of Dada. Is it any wonder that the rejection of reason made manifest by senseless atrocities should lead to the embrace of irrationality and disorder among the West's subaltern artistic communities? Marcel Janco, one of the first Dadaists, cited this rationale: "We had lost confidence in our culture. Everything had to be demolished. We would begin again after the *tabula rasa*." Thus, we find the overturning of what was once considered art: a urinal becomes the *Fountain* after Marcel Duchamp signs it "R. Mutt" in 1917, the nonsense syllables of Hugo Ball and Kurt Schwitters transform into "sound poems," and dancers in cardboard cubist costumes accompanied by foghorns and typewriters metamorphosize into the ballet *Parade*. Unsurprisingly, many commentators, including founding members, have described Dada as an "anti-art" movement. Notwithstanding such a designation, Dadaism has left a lasting imprint on modern Western art.

Example:

4. According to some estimates, prior to the beginning of World War I in 1914, more than four-fifths of the world's landmass was controlled by European nations or former colonies such as the United States. If this figure is accurate, what effect does it have on the passage?
    A. It bolsters the author's suggestion that European colonialism was an overbearing force.
    B. It weakens the author's assertion that World War I was instigated by provincial disputes.
    C. It strengthens the author's claim that Europe is the only place in the world that mattered.
    D. It challenges the author's insinuation that European rulers ignored the rest of the globe.

Solution: With lengthy opening details, you'll want to skip right to the question itself, which asks for the *effect* on the passage. Assess that this is a Strengthen–Weaken (Beyond) question and that your task will be to determine the relevance of this new data. As you read the stem a second time, now closely, think about how the information presented either supports or challenges statements from the passage. The evidence provided has nothing to do directly with the Dada movement, which doesn't even start until after the commencement of the war, but it does pertain to the author's discussion of Europe and World War I at the end of the first paragraph, so Executing your Plan requires rereading that portion of the text to see whether anything there would be impacted. If nothing else, the figure cited does seem to support the author's opinionated characterization of *the dominating culture of colonialist Europe*.

If you look for a match, you'll find that **(A)** works perfectly. The word *overbearing* is just a synonym for *dominating*, so there is a close connection with language actually used in the passage. On Test Day, you would select this answer and then move on to the next question without even bothering to read the wrong answers.

For our purposes, however, it's worth reviewing where the others go wrong. **(B)** may point to a claim that the author actually makes, but the fact from the question stem does not really pose a threat to it. Regardless of how much landmass each country controlled, the catalyst for World War I could remain disputes among these countries. This answer is a Faulty Use of Detail. In contrast, **(C)** is wrong

MCAT Critical Analysis and Reasoning Skills

for two reasons: first, it's Out of Scope because this is not something the author endorses; the phrase *the European subcontinent apparently being the only part of the world that mattered* is used almost sarcastically when the author points out that "World" War I actually took place exclusively in Europe. Second, even if the author did have this view, the statement is a value judgment—a matter of opinion—which cannot be directly affected by geographic facts. Finally, **(D)** does contain a factual claim that would be challenged by the question stem. The only problem is that this is not an insinuation of the author, who explicitly refers to the rulers *turn[ing] inward* after *violent exploitation of other peoples*, suggesting they were indeed gazing outward before the start of the war, the time period noted in the question stem. It's also Out of Scope.

Example:

5. Which of the following, if true, would most threaten what the author says in the final sentence of the second paragraph?
    A. A large majority of members of the general public, when asked to identify the most important work of art of the 1900s, fail to mention an example from the Dada movement.
    B. Other prominent 20th-century artistic movements, such as Surrealism and Pop Art, were also commonly described as "anti-art" by their most influential participants.
    C. The consensus among art historians today is that Dada was merely a brief departure from the principal themes in European art that evolved during the 20th century.
    D. Some of the founding members of the Dada movement were sympathetic to the radical view that, far from being anti-art, Dada was the purest form of art imaginable.

Solution: Although this question is perhaps more difficult to Assess than the previous example, the hypothetical *if true* is strong evidence that the answer choices will be new elements and that this is indeed a Strengthen–Weaken (Beyond) question. Clearly the task is to Weaken the author's claim; *threaten* is just a less common way of saying this.

Plan to follow the reference in the stem and reread that last line, as well as the preceding one, to gain additional context. The penultimate sentence (the one before the last) suggests that some people think of Dada as *anti-art*, but the last one contests this by noting Dada's *lasting imprint on modern Western art*. Thus, we are most likely looking for a refutation of the author's idea that Dada was influential.

There is a chance, however, that the correct answer challenges the author's rejection of the term *anti-art*. In other words, if we cancel out the confusing double negatives, the correct answer could be one that supports the idea that Dada is anti-art.

An elaborate prediction, such as the one just formulated, will usually result in a quicker match. In this case, we find it in **(C)**. An appeal to expert opinion is often an acceptable form of evidence for arts and literary passages, as can be seen here in the author's name-dropping and quotation-citing, so consensus among experts should carry even more weight. The scholars' idea of a *brief departure* clashes directly with the *lasting imprint* from the text, so we can be confident in identifying **(C)** as correct.

You would ordinarily want to move on to the next question at this point, but it's worth learning from the flaws in the wrong options. **(A)** clearly does not support what the author says, but it poses less of a challenge than it may seem to. For one, popular opinion could easily be mistaken about how influential particular works of art are. But, more substantially, leaving a *lasting imprint* is not identical to being the singularly most important movement of the century. Thus, the threat presented by **(A)** is extremely weak at best. The remaining options are Opposites that actually support the author. **(B)** is completely consistent with the passage, strengthening the point that simply calling a movement *anti-art* does not necessarily make it so. **(D)** has no impact on the passage. Even if some of the *founding members* of Dadaism did not think of the movement as being *anti-art*, many of the founding members certainly could have held this view. Further, this answer choice attempts to weaken the penultimate sentence—not the last one of the paragraph.

MCAT Expertise

Appeals to authority or expert opinion are common in CARS passages, particularly those involving disciplines like the arts or literature, in which value judgments and other opinions play a prominent role. Widespread agreement among experts, though rare in these fields, would of course provide even stronger support. That said, be mindful of the field that you are reading about and the types of argumentation that the author chooses to employ. Quoting experts may provide decent support for a passage on an artistic movement, but such testimony will carry far less weight in more empirical social sciences like psychology or economics. Non-expert opinions tend to carry even less weight: while authors may occasionally draw on popular opinion to support arguments, actual surveys of public opinion are seldom seen outside of a small number of cases, confined primarily to political science.

11.3 Other *Reasoning Beyond the Text* Questions

Some question tasks that transcend the passage fall into neither the Strengthen–Weaken (Beyond) nor Apply types. Because none of these occurs especially frequently, we will simply discuss three cases that have made an appearance on past MCAT exams.

PROBABLE HYPOTHESIS

In many ways, **Probable Hypothesis questions** are a counterpart to Apply questions but instead of asking about the Outcomes of the new situations they present, they ask about the likely causes. After presenting the new details, these stems will

MCAT Expertise

According to our research of released AAMC material, there is usually one *Reasoning Beyond the Text* question that is neither an Apply nor a Strengthen–Weaken (Beyond the Passage) question.

ask for *a probable hypothesis*, *the likely cause*, or *the most reasonable explanation based on the passage*. Working backward from a given effect to its probable cause can be more difficult, but again your task will be to use the Logic keywords that reveal analogous cause–effect relationships in the passage to form your prediction. If no match can be found, eliminate any answers that contradict claims the author states or suggests elsewhere in the passage.

ALTERNATIVE EXPLANATION

Alternative Explanation questions also concern causes. However, what makes these questions tricky is that they start with a phenomenon that might be directly from the passage but ask for a cause that is not given and that may not even be very similar to anything discussed. It is almost impossible to predict the answers to such questions, so Plan to go through each answer choice, eliminating any that would not produce the result given in the stem. If stuck between multiple answers that seem just about equally likely to serve as the cause, eliminate those that would most conflict with other parts of the passage. A correct alternative explanation won't be one that the author provides, but, other than departing from the author's original explanation, it should not significantly contradict the author.

PASSAGE ALTERATION

One other less common *Reasoning Beyond the Text* question type will inquire about changes that the author could make to the passage to allow it to be consistent with new information provided. These are appropriately called **Passage Alteration questions**. In many ways, these are like the rare instances of *Reasoning Within the Text* that require resolving paradoxes, except that they will include some new information that contradicts what the author says or implies. The correct answer to these questions will typically be the one that produces the desired effect with the *least* amount of modification to ideas originally presented in the passage.

Conclusion

And so this brings us to the end of our discussion of question types. If you still find yourself confused about which name corresponds to which type, don't worry! The common stems, tailored strategies, and worked examples of the last three chapters are designed to be only your first exposure to the intricacies of the Kaplan Method for CARS questions, the question types and tasks, and the Wrong Answer

Pathologies. Continue practicing until the Method becomes second nature for you. In the end, it will be less important to be able to simply name the question type than to know what to do with such a question. The number one best way to improve your performance with CARS questions and get that higher score you deserve is practice accompanied by effective review—the subject of our final chapter.

MCAT Critical Analysis and Reasoning Skills

CONCEPT AND STRATEGY SUMMARY

Apply Questions

- Assess: **Apply questions** require you to take the information given in the passage and extrapolate it to a new context. Apply questions may ask for one of three tasks.
    - They may ask for the author's **Response** to a situation, using words like *response*, *reply*, *most likely to agree with*, or *least consistent with*.
    - They may ask for the most probable **Outcome** in a situation, using words like *outcome*, *result*, *expectation*, or *consequence*.
    - They may ask for an **Example** of an idea discussed in the passage, using words like *example* or *instance*.
    - These questions often begin with words like *Suppose*, *Consider*, or *Imagine*.
- Plan: If the question stem is long, jump to the end to determine what it's asking. Read any information given in the question stem closely, looking for hints that connect it to the passage.
- Execute: Reread the relevant text, keeping in mind the specific type of Apply question involved.
    - For Response questions, determine the author's key beliefs, which are generally reflected in the passage using Author keywords.
    - For Outcome questions, pay attention to cause–effect relationships in the passage, which are generally reflected in the passage using Logic keywords.
    - For Example questions, look for text that provides definitions, explanations, or the author's own example, noting any necessary or sufficient conditions.
- Answer: Match your expectations with the right answer. If there is no clear match, or if you cannot perform any of the earlier steps of the Kaplan Method for CARS Questions, use process of elimination.
    - Eliminate any answer choices that are inconsistent with the author's views, especially for Response questions.
    - Eliminate any answer choice that does not contain necessary conditions (which must occur in all instances of a concept), especially for Example questions.

Strengthen–Weaken (Beyond the Passage) Questions
- Assess: **Strengthen–Weaken (Beyond the Passage) questions** concern the logical relationship between conclusions and the evidence that strengthens them or the refutations that weaken them.
  - These questions often contain words like *relate*, *support*, *challenge*, *relevance*, *significance*, or *impact*. In contrast to Strengthen–Weaken (Within the Passage) questions, they often contain words like *could* or *would*.
  - These questions are closely related to Strengthen–Weaken (Within the Passage) questions, which use information directly from the passage, rather than bringing in a new piece of information.
  - Read the question stem closely, looking for hints of analogy to parts of the passage.
- Plan: Determine the two claims and the connection between them; you will usually be given at least one of these elements and will have to find the other(s).
  - Identify where each piece of the argument can be found: in the question stem, in the passage, or in the answer choices.
  - If no claims are given in the question stem, plan to triage it and answer it by process of elimination later.
  - If one claim is given in the question stem, determine if it is a conclusion, a piece of evidence, or a refutation.
  - If two claims are given in the question stem, identify the relationship between them.
- Execute: Research the relevant text to determine the missing claim or the connection between them. Use Logic keywords to help assemble the argument.
- Answer: Match your expectations with the right answer. If there is no clear match, or if you cannot perform any of the earlier steps of the Kaplan Method for CARS Questions, use process of elimination.

Other *Reasoning Beyond the Text* Questions
- **Probable Hypothesis questions** ask for causes of new situations presented in the question stem.
  - These questions often contain words like *probable hypothesis*, *likely cause*, or *most reasonable explanation*.
  - Approach these questions like Apply questions, except that you are looking for analogous cause–effect relationships in the passage.

- **Alternative Explanation questions** ask for causes that differ from the ones given in the passage but that still provide an explanation for a phenomenon.
    - These questions often contain words like *alternative explanation*, *other cause*, or *different reason*.
    - Approach these questions by eliminating any answer choice that would not lead to the effect in the question stem. If stuck between multiple answers, eliminate those that conflict most significantly with the passage.
- **Passage Alteration questions** ask for changes the author could make to the passage to make it consistent with new information.
    - These questions often contain words like *alter*, *change*, or *update*.
    - Approach these questions by looking for the answer that produces the desired effect with the least amount of modification to the ideas in the passage.

11: Question Types III

WORKED EXAMPLE

Use the Worked Example below, in tandem with the subsequent practice passages, to internalize and apply the strategies described in this chapter. The Worked Example matches the specifications and style of a typical MCAT *Critical Analysis and Reasoning Skills* (CARS) passage.

Passage	Analysis
Summer to winter to summer yet again, morning to night and then dawn once more. All things in life seem to cycle, and so too do trends in art. Styles, of course, do coexist and always have. Life is rarely as neatly divided between night and day as we might wish. But throughout history, art has followed one main avenue and then reversed direction time and time again, thus producing the classicism *vs.* romanticism (or expressionism) dichotomy.	The passage begins with vivid imagery that is representative of *cycle[s]*. The author then reaches the thesis of the passage: *trends in art* also cycle. The phrase *of course* indicates that we should focus on what is novel rather than what is already known: we are less interested in the possibility of coexistence and rather will focus on the *dichotomy* between *classicism and romanticism (or expressionism)*. A Label for this paragraph might be: **P1.** Classicism alternates with romanticism (expressionism)
Classicism in art primarily refers to clean, cool imagery. In the High Classical period in ancient Greece, idealized sculpted figures of young men and women were perfect in proportion, the picture of health and vitality. The subsequent ancient Greek Hellenistic art swept in a more expressive era in which figures depicted actual people, with an emphasis on their individuality. Eschewing the sleek lines of the Classical period, sculptors lent their images a sense of weight so that clothing and hair looked a bit waterlogged. However, this additional substance often produces a sense of expressive motion. The goddess Nike (Victory) of Samothrace races forward as her windswept drapery creates wet wings behind her. The rational distance of the earlier Classical sculptures has given way to expressions that convey a more passionate, romantic essence.	We get definitions for the terms in the dichotomy: *classicism* is associated with *clean, cool imagery* and *rational distance* whereas *expressive* art *emphasi[zes] individuality* and has images with more *weight*, more *expressive motion*, and *a more passionate, romantic essence*. This dichotomy is shown between two art styles in ancient Greece: *the High Classical period* and *Hellenistic art*. This could be Labeled: **P2.** Examples and characteristics of classicism (Greek High Classical) and expressionism (Hellenistic art)

MCAT Critical Analysis and Reasoning Skills

Passage	Analysis
Not surprisingly, the French Neoclassicists, from about 1750 to 1850, looked to the Classical age for inspiration. Painting and sculpture contained the same refined, dignified qualities as the earlier work, although employing contemporary subjects. Portraits of both aristocrats and commoners reveal the late 18th- to early 19th-century "re-vision" of Classical times in everything from fashion to furniture and architecture. It was the Romanticists, though, who put the soul back into art. Their technique was looser, emitting the sense that the artist's hand had just lifted off the canvas or finished chiseling the stone. Brooding compositions described exotic locales in the Middle and Far East. Heroic stories detailed contemporary shipwrecks, battles, and civilian revolutions. Neoclassical works taste of buttered toast where romantic pieces taste of hot spice.	*French Neoclassicists* are *1750 to 1850* classicists, and our author sees them as very similar to the art of the *Classical age* but with *contemporary subjects*. Then, Romantic pieces *put the soul back into art* with a *looser* style and *exotic* and *heroic* themes. The author uses another sensory modality, *taste*, to further get across this dichotomy. A Label for this paragraph is: **P3.** French Neoclassicists = contemporary classical; Romanticists = soulful, exotic, heroic art
Classicism evolved into two camps during the 20th century. A realism trend continued, in which artists depicted the world along the lines of human perception. The Regionalists in the early part of the century reflected life in America's backcountry. Grant Wood's 1930 painting "American Gothic" presents a no-nonsense farm couple staring the viewer straight in the eye. They exude the basic goodness and solidity of their nature. They stand together for eternity, more as emblems of an age and ideal than true individuals. Interestingly, the same cool distance resulted later on in the abstraction of Minimalism, beginning around the early 1960s. Minimalist artists created no figures or references to the outside world. Instead, the sharp edges of their geometric shapes, unmixed colors, and lack of visible brush or carving stroke embody the same distilled, classical calm.	Another set of two items is given at the beginning of the paragraph; in this case, *two camps* of classicism in the *20th century*. The first is *realism*, which includes the *Regionalists* and the example of *American Gothic*. The second is *Minimalism*, which comes later. Characteristics specific to each style are mentioned—*reflect[ing] life in America's backcountry* for realism and *abstraction…geometric shapes, unmixed colors, and lack of visible* techniques in Minimalism—as are characteristics that define both of these as classical style: *cool distance* and *distilled, classical calm*. Given that the author keeps alternating chronologically between classical and romantic styles throughout the passage, we expect the next paragraph to address recent romantic styles. This paragraph's Label could be: **P4.** 20th-century classicism: realism (Regionalists) = American life; Minimalism = abstract, geometric, unmixed colors

11: Question Types III

Passage	Analysis
Between these two periods, America birthed Abstract Expressionism, its most fervent art form. Painters abandoned realistic, figurative images and thrust their inner emotions or the invisible vibrations of the universe onto canvas. Virtuous brushwork and color flash across flat surfaces with a magnetism and energy unknown before. These huge compositions take your breath away. Abstract Expressionism, the nation's first unique art movement, exudes all the brashness of a young upstart, even as the more classically oriented works that bracket this movement recall an esteemed, stately heritage.	We are told that the romantic trend of *Abstract Expressionism* actually happened between the two classical ones from the previous paragraph. This style is characterized by depicting *inner emotions or the invisible vibrations of the universe* with *magnetism and energy*. The author's tone is even more emotive, mentioning that the works in this style *take your breath away*. At the end, this *brash* movement is contrasted with the *esteemed, stately* ones that precede and follow it. A Label for this paragraph is: **P5.** Abstract Expressionism = emotions and energy; between movements in P4
The time between the cycle from cool to expressive, ideal to romantic has shortened of late. The current art scene hosts art from both sides now. Perhaps the future will bring us more artists who freely sew the two ends of the continuum together.	Here's our dichotomy: *cool* and *ideal* (classicism) *vs.* *expressive* and *romantic*, and we find out that they are alternating more quickly. The author suggests that the two styles may be brought together at some point in the future. Our Label for this paragraph is: **P6.** Cycle time shortening, future artists draw from both?

Here's a sample Outline and Goal for this passage:

P1. Classicism alternates with romanticism (expressionism)

P2. Examples and characteristics of classicism (Greek High Classical) and expressionism (Hellenistic art)

P3. French Neoclassicists = contemporary classical; romanticists = soulful, exotic, heroic art

P4. 20th-century classicism: realism (Regionalists) = American life; Minimalism = abstract, geometric, unmixed colors

P5. Abstract Expressionism = emotions and energy; between movements in P4

P6. Cycle time shortening, future artists draw from both?

Goal: To chronicle the historical cycling between classicism and romanticism in art styles and suggest the dichotomy may be lessening

MCAT Critical Analysis and Reasoning Skills

Question	Analysis
1. Assume that most people at the time said the work of the Regionalists was emotionally overwhelming. What effect would this have on the author's arguments?	Because this question brings in new information and asks about its *effect*, this is a Strengthen–Weaken (Beyond the Passage) question. Regionalism supposedly evolved from classicism, but *emotionally overwhelming* is a much better descriptor for Abstract Expressionism, as described in paragraph 5: these artists *thrust their inner emotions...onto canvas, creating huge compositions [that] take your breath away*. Because the new information in the question stem seems to go against the author, we Plan to look at the answer choices that include the word *weaken*.
A. It would strengthen the assertion that the time period between styles is currently shortening.	**(A)** and **(B)** can be eliminated immediately because they say *strengthen*.
B. It would strengthen the claim that classicism primarily refers to clean, cool, imagery.	
C. It would weaken the claim that art styles can coexist.	Because the new information implies that Regionalists have some expressive attributes, the claim that classicism and romanticism can coexist is strengthened, not weakened—making **(C)** an Opposite answer.
D. It would weaken the claim that Regionalism was a type of classicism.	The new evidence suggests that Regionalism could have been misclassified as classicism when it has expressive aspects, confirming **(D)** as the correct answer.

11: Question Types III

Question	Analysis
2. In 1801, a French Neoclassicist announced that "I seek to infuse the modern era with the historical weight of a great past." On the evidence of the passage, he could have best achieved his goal by producing:	This is an Apply question of the Example subtype. Given the Roman numerals, it is a Scattered Apply question. As a French Neoclassicist, this artist would aim to have the *refined, dignified qualities* of classicism, *although employing contemporary subjects*. The quote in the stem also indicates that a reference to something *historical* or the *past* should also appear in the answer choice.
I. a statue of Napoleon Bonaparte in which he is costumed plainly as an ancient Greek emperor.	In Statement I, we see Napoleon is the contemporary subject matter but the artist is inspired by ancient times and uses a plain costume, reminiscent of classical style. This fits the criteria, so Statement I must appear in the correct answer.
II. a full-length portrait of Romantic Novelist Victor Hugo resembling one of his heroic characters.	In Statement II, representing Hugo as a hero makes no connection to the past—*refined* and *dignified* or otherwise. This falls short of what we need, so Statement II must not appear in the correct answer.
III. a painting of Queen Marie Antoinette in which she is indistinguishable from the courtiers who surround her.	Statement III similarly does not make a connection to the past, and therefore must not appear in the correct answer. Further, to make the Queen blend in with her court would be the opposite of making her appear *refined* and *dignified*.
A. I only	**(A)** contains the correct Statement and is the right answer.
B. III only C. I and II only D. II and III only	**(B)**, **(C)**, and **(D)** contain incorrect Statements.

MCAT Critical Analysis and Reasoning Skills

Question	Analysis
3. Which of the following musical experiences is most analogous to Minimalism?	This is an Apply question asking for a musical Example that is similar to Minimalism in art. Minimalism is described at the end of paragraph 4, and all of the characteristics of this style are said to *embody...distilled, classical calm* and is a form of classicism that involves abstraction and also ideals like calm.
A. Agitated classical music in a large concert hall	**(A)** intentionally uses the word *classical* as a Faulty Use of Detail, but the word *agitated* indicates that this music is anything but calm. Eliminate this answer choice.
B. Electronic elevator music playing quietly	Elevators are typically calm places—the phrase *elevator music* even tends to have the connotation of innocuous or boring, implying that this music is quite calm. This makes **(B)** the correct answer.
C. Hard rock music blasting through speakers	We can rule out **(C)** because *hard rock* and *blasting* indicate that this music anything but calm.
D. A repetitive tape loop of country music	A repetitive tape loop would likely be irksome and not particularly calm, making **(D)** incorrect as well. This answer choice is a good trap if you are familiar with music history as many Minimalist composers did indeed employ tape loops in their music; however, this answer is not supported by the passage and is therefore incorrect.

11: Question Types III

Question	Analysis
4. With which statement would the author most likely DISAGREE?	This is an Inference question of the Implication subtype; we are looking for something that the author would not agree with. It is hard to come up with a solid prediction for this question, so Plan to go through systematically and eliminate any choices that are consistent with the author's opinions or that are Out of Scope.
A. Late 18th- to early 19th-century French Romantic art had a sense of personality that Neoclassicism lacked.	**(A)** clearly addresses the dichotomy we are getting comfortable with: classical = cool, romantic = expressive. The author would agree that *Romantic art* has more personality than a classical movement, so eliminate this answer choice.
B. At any given time it can be difficult to pinpoint a strong dichotomy between prevailing styles and previous ones.	**(B)** reflects what the author tells us in the first paragraph: that *styles, of course, do coexist* and that *life is rarely as neatly divided...as we might wish*. This answer can be eliminated as well.
C. French Romanticism lacked an immediacy that was apparent in Neoclassical painting.	**(C)** goes against the author's description of classicism as having a coolness and distance to it, whereas romanticism has more expressive, emotional impact. Romanticism should have the immediacy that classicism lacks—this answer choice is an Opposite of the author's opinion and is therefore the correct answer with which the author would *DISAGREE*.
D. French Romanticism was the polar opposite of Neoclassicism.	This passage repeatedly draws a polar contrast between classicism and romanticism. **(D)** is a valid inference based on the passage and therefore can be eliminated.

MCAT Critical Analysis and Reasoning Skills

Question	Analysis
5. Suppose that most late 18th-century French drawings are exotic and exciting. Which passage assertion would be most WEAKENED?	This is another Strengthen–Weaken (Beyond the Passage) question that provides new evidence and asks us to identify the conclusion that is most *WEAKENED* by it. The *late 18th-century* is a reference to the time period given at the beginning of the third paragraph, *1750 to 1850*. This is the French Neoclassicist era—a classical period—but *exotic* and *exciting* are descriptors of romanticism. This implies that many artists were not actually using classical themes at the time.
A. Classicism was more popular than Romanticism in late 18th-century France.	**(A)** uses comparative language, but how *popular* one style is in comparison to another is not addressed at all in the passage and so this choice can be eliminated.
B. The Neoclassicists' inspiration had pervasive effects.	This claim is made in paragraph 3: according to the author, Neoclassicism impacted *everything from fashion to furniture and architecture*. However, because we were surprised to hear about romanticism in a classical period, it no longer seems that the Neoclassicists were as *pervasive* as the author described. Therefore, **(B)** is the correct answer.
C. Inner emotions can be imbued into artwork.	**(C)** is certainly a valid inference, but it is not affected in any way by the question stem's information.
D. Neoclassical works are less dramatic than works of the Classical age.	**(D)** also makes a comparison that was never mentioned explicitly in the passage—the author never addresses how *dramatic* one classical style is compared to another classical style.

11: Question Types III

Question	Analysis
6. Suppose that romantic artists and classically-oriented artists began borrowing heavily from one another. This finding would support the view that:	This is a Strengthen–Weaken (Beyond the Passage) question, based on the words *Suppose* and *would support the view*. The new evidence in the question stem is consistent with the final point the author makes—that these two styles may eventually be *sew[n]...together*. To Execute our Plan, we Scan for an answer choice that is consistent with the Goal of the passage.
A. romanticism and classicism are independent movements.	**(A)** contradicts the author's point in the last paragraph. Further, the question stem shows a case where the two movements are converging—not that they are *independent* of each other. Eliminate this answer choice.
B. the past inevitably influences the future.	**(B)** uses Extreme language and was never overtly claimed by the author. While the author provides some examples of the *past...influenc[ing] the future*, such as French Neoclassicism, the word *inevitably* is far too strong, as many movements that are not strongly influenced by the past, such as Abstract Expressionism, are also described.
C. the cycle from cool to expressive art styles is shortening.	**(C)** might be tempting because it is a claim the author makes. However, *borrowing heavily* from each other at one point in time would mean there wouldn't be two distinct styles that are cycling.
D. artists might be beginning to make less of a distinction between these two schools of art.	**(D)** uses the Moderating keyword *might* and correctly states the author's main point at the end of the last paragraph. That's our match.

Question	Analysis
7. Which of the following, if true, would constitute a reason Neoclassicists looked to Classical Greece for inspiration?	For this Strengthen–Weaken (Beyond the Passage) question, we are looking for a reason Neoclassicists would look to the High Classicism style in ancient Greece for inspiration. It is challenging to make a prediction, but we should look at the answers with an eye toward the themes the author has identified as part of the classical mode.
A. Nineteenth-century French nobility admired the ancient period because it was one in which even ordinary citizens acquired important art.	**(A)** does not address the idea of classicism at all; further, there is no obvious reason why *nobility* would like an art style that *even ordinary citizens* could acquire. Eliminate this answer choice.
B. Neoclassical painters and architects were impressed by the wide range of human feelings that Classical Greek sculptors captured in their marble works.	**(B)** discusses *feelings*, which the passage indicates are actually the hallmark of the opposite art style: expressionism. Eliminate this Opposite answer choice.
C. Nineteenth-century French elite idealized the cultural expressions of the ancient past as conveying regal grandeur, devoid of intense emotions.	**(C)** references the absence of emotions, which matches our understanding of classical values. In addition, it makes sense that the *French elite* would be attracted to art that *convey[ed] regal grandeur*. This is the correct answer choice.
D. Neoclassicists found that appealing to patrons' interest in antiquity allowed them to move their style toward the expressionist ideal.	**(D)** implies that Neoclassicists desired to move their style out of the classical mode and into the expressionist mode, but there is no evidence in the passage that these artists desired to do so.

Practice Questions

Passage 1 (Questions 1–5)

Post-structuralist literary criticism was developed largely in reaction to Saussurian linguistic theory, which first expressed the relationship between words and the concepts they denote. In Saussurian linguistics, an actual word is referred to as a "signifier"—the "sound image" made by the word "train," for instance, constitutes a signifier. At the same time, the idea evoked by the signifier is termed a "signified." Saussure argued that the structural relationship between a signifier and a signified constituted a "linguistic sign." He saw language as made up entirely of such signs, or structural relationships, and argued that the relationship that constituted these signs was actually arbitrary and based on common usage rather than on some necessary link. He did believe, however, that certain "signifiers" (words) could be permanently linked to specific "signifieds" (concepts) in order to create stable, predictable relationships that evoked constant meanings.

In contrast to Saussurian linguistics, the post-structuralist view contends that there exists no system of describing ourselves, or of communicating with one another, which does not somehow use our indigenous language systems. To post-structuralists, language defines our identities and is required if we are to maintain those identities. In this view of language, any signifier always signifies another signifier. Definitions and meanings always take the form of metaphors: one term can only be defined as being another term. To change the meaning of a term, one must only change the metaphor through which that term is defined. Meaning shifts from one signifier to another, and because of this, no act of signification is ever fully closed or fully complete.

Because the post-structuralists do not view necessary connections as composing permanent linguistic sign relationships, they reject the idea of absolute meaning. Because language constantly shifts along a chain of meaning, "absolute" meanings cannot exist. Language in this view can never be viewed as entirely stable. Whereas Saussure believed that linguistic sign relationships could create stable, consistent meanings between terms and the images they evoke, post-structuralists argue that meaning can be established only through discourse. Thus, meaning is never absolute, immutable, or concrete because it is always dependent upon the differing and constantly shifting discourse in which language terms operate.

It is here that the divergence of these two schools of thought becomes readily apparent, insofar as they concern themselves with fields beyond linguistics. The implications of the nearly irrefutable, albeit bleak, reasoning that is so fundamental to post-structural thought reach far beyond the confines of linguistics. The idea of conceptual instability is a manifestation of the existential phenomenology that heavily influenced the work of innumerable scholars in disparate fields through the 20th century. Where structuralism was grounded in linguistics and made inroads into the human sciences, the very ideas that presuppose post-structuralism are rooted more ambitiously in the central discussion of human nature. It is for this reason that existentialism and post-structuralism continue to flourish and have inserted themselves into our enduring understanding of what it is to be human while structuralism and Saussurian linguistics hold a devoted place in scarcely few discussions beyond linguistic relativism.

MCAT Critical Analysis and Reasoning Skills

1. Which of the following best adheres to the post-structuralist view of meaning as presented in the passage?

   A. The words "signifier" and "signified" are expressed in different ways across several languages.
   B. Each member of a literary club puts forth a unique interpretation of a fairy tale based on his or her academic background.
   C. The relationship between a father and his son shifts dramatically during the course of a novel.
   D. All ten members of a focus group derive the same meaning from the preview of a new situation comedy.

2. In addressing a class, a professor describes the mind first as a blank slate to be written on and later as a garden to be cultivated. In light of the information in the passage, the professor's method best represents:

   A. the immutable nature of meaning.
   B. a Saussurian relationship between words and the concepts they denote.
   C. the arbitrary use of language in academia.
   D. a post-structuralist change in the meaning of a term.

3. "Broad, open-ended study is frivolous and indulgent. It is through deliberate, focused study that knowledge advances." How does this statement affect the author's argument in paragraph 4?

   A. It weakens the author's argument.
   B. It strengthens the author's argument.
   C. It neither strengthens nor weakens the author's argument.
   D. It could both strengthen and weaken different parts of the author's argument.

4. A study finds that humans taught a fabricated word and its definition immediately form an association between that word and its definition. After repeated exposure to the word in different contexts, the same people consistently offered different definitions for the word than the one they were initially given. This strengthens:

   A. post-structural theory because the initial signifier did not maintain its association with the initial signified.
   B. post-structural theory because the initial signifier was most likely understood through a metaphor that changed.
   C. Saussurian linguistic theory because of the initial association of signifier and signified.
   D. neither theory, either theory, or both theories; more information on the underlying mechanism that precipitated the change is needed.

5. Which of the following examples is LEAST analogous to the Saussurian understanding of linguistic relativism?

   A. A dog's response to the "sit" command could not be changed upon retraining, though its response to the "stop" command could be changed.
   B. A study reveals that words learned in any language stimulate the same neurons regardless of attempts to retrain the meaning of these words over several years.
   C. The definition of the word "cool" has evolved drastically over the years depending upon social context and common usage.
   D. A study revealed that the presentation of a word to a listener immediately resulted in a localized brain region becoming activated before the listener could offer the word's definition.

Passage 2 (Questions 6–10)

Where once they were surrounded and protected by vast wilderness, many of America's national parks are now adversely affected by activities occurring outside their boundaries. The National Park Organic Act established the national park system and empowered the Secretary of the Interior to manage activities within the parks. According to this act, however, conditions outside park boundaries are not subject to regulation by the Park Service unless they involve the direct use of park resources.

Several approaches to protecting the national parks from external degradation have been proposed, such as focusing on enacting federal legislation granting the National Park Service broader powers over lands adjacent to the national parks. Legislation addressing external threats to the national parks twice passed the House of Representatives but died without action in the Senate. Giving the states bordering the parks a significant and meaningful role in developing federal park management policy was also brought to the table as a possible remedy, though this solution is rife with its own problems.

Because the livelihood of many citizens is linked to the management of national parks, local politicians often encourage state involvement in federal planning. In spite of this, current state legislation has been an ineffective legal solution to the dangers facing park wildlife and the parks themselves. For instance, state legislatures have not effectively addressed the fundamental policy issue of whether states should be responsible for protecting park wildlife. State land use and environmental statutes, moreover, are often not intended to solve problems of this nature. Timber harvesting, ranching, and energy exploration compete with wildlife within the local ecosystem while priorities among different land uses are not generally established by current legislation. Additionally, often no mechanism exists to coordinate planning by the state environmental regulatory agencies, thus limiting the impact of legislation aimed at protecting park wildlife and the larger park ecosystem.

Even if these deficiencies can be overcome, state participation must be consistent with existing federal legislation. States lack jurisdiction within national parks themselves, and therefore state solutions cannot reach activities inside the parks, thus limiting state action to the land adjacent to the national parks. Most of this land falls under federal oversight by default due to ownership laws. Under the supremacy clause, federal laws and regulations supersede state action if state law conflicts with federal legislation, if Congress precludes local regulation, or if federal regulation is so pervasive that no room remains for state control. With the assumption that federal regulations leave open the possibility of state control, state participation in policy-making must be harmonized with existing federal legislation.

The residents of states bordering national parks are affected by park management policies as well. They in turn affect the success of those policies. This interrelationship must be considered in responding to the external threats problem. Local participation is necessary in deciding how to protect park wildlife. Local interests should not, however, dictate national policy, nor should they be used as pretext to ignore the threats to national parks or the regions that surround them.

6. Suppose that a state government's environmental policies were contributing to the gradual extinction of a species in a national park within its territory. How would this information affect the author's argument?

   A. It would strengthen the author's argument.
   B. It would weaken the author's argument.
   C. It would neither strengthen nor weaken the author's argument.
   D. It would strengthen the author's argument only if it were shown that the National Park Service was not trying to save this species.

7. The author of the passage would most strongly believe that which of the following legislative issues where federal and state policy are in conflict can be effectively overcome?

   A. A state seeking to legalize capital punishment does not entirely comply with federal regulations but is allowed to revise their proposal.
   B. A state seeking to protect a species, including its habitat within a major national park, appeals to the supremacy clause to pass its legislation in its entirety.
   C. A national park seeking legislation pertaining to the misuse of nearby municipal lands appeals first to the state legislature which proposes a thorough review of federal precedent.
   D. A state seeking to legalize physician-assisted suicide has given thorough consideration to the relevant federal statutes and existing legal precedent in drafting the proposed legislation.

8. In light of the viewpoints presented throughout the passage, the author would most likely support which of the following actions by state governments?

   A. Efforts to buy back land adjacent to national parks from the federal government
   B. Efforts to force the federal government to give up control of national parks to states
   C. Efforts to curb activities that are harmful to national park wildlife
   D. Efforts to reduce the power of the Secretary of the Interior

9. Which of the following situations is analogous to the challenge facing national parks, as outlined by the author?

   I. A man trying to complete his work in his own office is having difficulty doing so due to the noise emanating from the office next to his.
   II. A state-run park cannot influence the process of the construction of a hydroelectric dam outside the park, on a river that the park is responsible for conserving.
   III. A municipality can regulate the release of toxic pollutants outside its boundaries only if it is funding the decontamination process.

   A. I and II only
   B. I and III only
   C. II and III only
   D. I, II, and III

10. Which of the following, if true, would most WEAKEN the author's main contention in Paragraph 4?

    A. There are examples of states having overcome minor legislative discrepancies with the federal legislature after lengthy and extremely expensive appeal processes.
    B. Under the National Park Organic Act, states are capable of appealing any decision supported by the supremacy clause where they pertain to national parks.
    C. No state has ever passed legislation that conflicts with federal regulations pertaining to national parks.
    D. The National Park Service experiences a revision of its mandate every three years.

Explanations to Practice Questions

Passage 1 (Questions 1–5)

Sample Passage Outline

P1. Saussure: arbitrary signifier (word) relation with signified (concept) = linguistic sign, can become concrete

P2. Post-structuralist: signifier defined only by other signifiers; understand through metaphors that can change

P3. Saussure = language can be stable, post-structuralist = language never stable, depends on discourse

P4. Post-structuralism endures because of broader ties to human nature; Saussurian theory restricted to linguistics

Goal: To explain the differences between Saussurian and post-structuralist theory

1. B
This question challenges us to Apply our understanding of the post-structuralist theory to the answer choices in order to find an Example. Paragraph 3 makes clear that the post-structuralists believe that *meaning is never absolute, immutable, or concrete*. Looking for the choice that best fits with this notion brings us to **(B)**. The fact that the *unique interpretations* are due to differences in *academic backgrounds* supports the notion that exposure to different fields of discourse leads to the differences in interpretation, which is consistent with the language of paragraph 3. **(A)** cleverly uses words from the passage in a new context. The fact that words are expressed differently in different languages is no knock against the absolute meaning of these words; they're expressed in various ways, as one would expect in different languages, but that doesn't mean they are defined differently. In **(C)**, the relationship between father and son may shift, but we'd have to see some variation in interpreting the *meaning* of this shift to get us into post-structuralist territory. As for **(D)**, a lack of consensus would seem more consistent with the post-structuralist view, although it's conceivable that the ten focus-groupers arrived at their opinions through discourse with each other. However, because such a state of affairs isn't indicated, we can't assume it.

2. D
In Application questions like this one, we must determine how the new Example relates to what's in the passage. In passages where the Goal is *to explain the differences between* two things, that usually means figuring out what camp the new situation falls into. The major task here is figuring out what the professor is doing, and how it relates to the theories in the passage. When the *professor describes the mind…as a blank slate*, or *as a garden*, she is using metaphors, which brings us into the post-structuralist camp. Moreover, she's shifting the metaphor, which relates precisely to the penultimate sentence of paragraph 2: *To change the meaning of a term, one must only change the metaphor through which that term is defined*. Therefore, the answer is **(D)**. **(A)** and **(B)** are Opposites because they presuppose constant meaning, whether it is described as immutable or Saussurian—Saussure posits *stable, predictable relationships that evoke constant meanings* according to the end of paragraph 1. **(C)** is Out of Scope as there is no mention of the *use of language in academia* in the passage.

3. C

This question provides new information and asks how it impacts the passage, making this a Strengthen–Weaken (Beyond the Passage) question. The new quotation suggests that widely applicable fields of study, such as post-structuralism as described in this paragraph, are frivolous and that narrowly-focused fields, such as Saussurian linguistics, advance knowledge. However, let's consider the author's argument in this paragraph. The author focuses on the differences in scope between the two camps and why post-structuralism is more enduring. He or she makes no mention of the advancement of knowledge or frivolity. The author may imply some partiality to post-structuralism, but there is no argument made regarding the focus of the new quotation. Thus, **(C)** is correct as this statement has no bearing on the author's argument.

4. D

This Strengthen–Weaken (Beyond the Passage) question challenges us to identify not only whether the new information *strengthens* post-structuralism or Saussurian theory, but also why. The first thing to recognize is that the study's results could support either theory. Even though Saussurian theory suggests that words *could be permanently linked* to concepts, it also recognizes that the relationship between a word and the concept it represents is *actually arbitrary and based on common usage rather than on some necessary link*. We also know that post-structuralism definitely supports changing definitions. Thus, we have to look at the reasoning in the answer choices to determine the correct answer. **(A)** may sound tempting because the reasoning sounds as if it's disproving Saussurian theory, but remember that Saussurian theory does account for signifieds changing depending upon *common usage* and, besides, Saussurian theory being proven incorrect wouldn't necessarily prove post-structuralism correct. **(B)** requires quite the leap in reasoning, featuring assumptions we just can't make—there is not enough support for it being a *metaphor that changed* that caused the change in meaning. **(C)** is also tempting because it mentions Saussurian theory and one of its tenets, although *the initial association of signifier and signified* does not really add any new information to strengthen the author's description. More importantly, this answer choice doesn't explain the all-important second half of the new information: why the association changed. Therefore, the answer must be **(D)**. This answer choice identifies the real problem with the other choices: we don't know the mechanism behind the change in association. Both theories account for a change in associations, so the mechanism by which the association changed needs to be described before we can draw any conclusions about which theory the study supports.

5. B

The challenge in this Application question is to transfer our knowledge of Saussurian linguistics to find three appropriate analogies or one choice that does not fit. This description, as noted in the Outline, appears in paragraph 1 of the passage where the author makes clear that signifiers are tied to signifieds in order to produce a sign, or linguistic understanding. This relationship can be, but is not necessarily, permanent. **(A)** adheres to this description well. The meaning can be permanently linked or changed, depending on the circumstances. **(B)** is too Extreme in its categorization of semantic links as permanent. The passage said that links are *arbitrary and based on common usage rather than on some necessary link*, so the categorization in this answer choice of all links being concrete is Extreme, and thus it is the correct answer. **(C)** fits in perfectly with the previous quotation that links are *based on common usage*. Finally, **(D)** is also a perfect explanation of Saussurian theory. The signifier is presented, it triggers brain activity, and the definition (signified) is provided.

Passage 1 (Questions 6–10)

Sample Passage Outline

P1. Problem: national parks affected by surrounding lands, but Park Service has little control

P2. Solutions tried: more power over neighboring lands (didn't pass), empower states to help

P3. Problems with state legislation as a solution

P4. State legislation must be consistent with federal policy

P5. Elements of local participation needed but should not be overriding (warnings)

Goal: To describe legislative issues in solutions for national park conservation

6. A

Here, the question stem asks us to reason how a novel situation *affect[s] the author's argument*, making this a Strengthen–Weaken (Beyond the Passage) question. First, we must consider what the author thinks about the effect of state governments on national parks. There's evidence throughout the passage, and especially in paragraph 3, that the author thinks that state governments' *land use and environmental statutes…compete with wildlife within the local ecosystems*. Therefore, the new example of state policies harming a species in a national park supports the author's argument. **(A)** is correct. **(B)** and **(C)** are, of course, necessarily incorrect by this logic. **(D)** throws the curveball of offering the same position—that this information *strengthens the author's argument*—but with a conditional qualifier: that *only if* statement. Whether or not the National Park Service is trying to save this species is irrelevant because the state policy still serves as an example of a state statute affecting wildlife in the ecosystem.

7. D

For this Apply question, we are asked to find which scenario would elicit a given Response from the author: which *legislative…conflict can be effectively overcome*. The author's argument can be found in the Outline's Label for paragraph 4, where it is argued that *state participation in policy-making must be harmonized with existing federal legislation*. This prediction should be sufficient to attack the answer choices. **(A)** offers a case of a state *not entirely comply[ing] with federal regulations*, which is the Opposite of what we're looking for. **(B)** again implies a case of a state at odds with the federal government, but the state *appeals to the supremacy clause* this time. Referring back to paragraph 4, we see that the supremacy clause dictates that federal statutes take precedence and thus the state will not even be successful by appealing to the supremacy clause. **(C)** gives the scenario of a federal service appealing to the state government about *municipal lands*. Not only do we know nothing about where municipal governments fall within this scheme, but also the passage gives no information about why the federal government would be appealing to the state government about something neither have any particular control over. If anything, the federal government would have much more say given the passage's description of the supremacy clause. **(D)** proposes a scenario where the state is seeking to align itself with federal requirements. This is the best match for our prediction and is the correct answer.

8. C

This is another Apply question of the Response subtype. We are looking for an answer *the author would most likely support*. First, let's consider what the author wants state governments to do. In general, the author is in favor of cooperation with the federal government and a greater effort to protect wildlife within the parks. Of these two priorities, only **(C)** offers a course of action that matches. **(A)** is a Faulty Use of Detail; states already own the *land adjacent to national parks*, so there would be no need to *buy back* this land. Both **(B)** and **(D)** would serve to limit the power of the federal government, which is never stated or implied by the passage.

9. B

For this Apply question, we must first identify what the author has deemed a *challenge facing national parks*. The opening paragraph states that *national parks are...adversely affected by activities occurring outside their boundaries* and that *conditions outside park boundaries are not subject to regulation...unless they involve the direct use of park resources*. These are the challenges that make way for the rest of the author's discussion. Therefore, we are looking for examples that are similar to these claims. Statement I is analogous to the above scenario because the *man trying to complete his work* is being adversely affected by conditions outside of his own space, much like the parks. **(C)** can be therefore eliminated. Statement II again offers a scenario where an entity is unable to control conditions outside its own space that would affect it. However, this time, the park should be able to influence proceedings according to the scenario described in the passage as *park resources* are committed to conserving the river. Therefore, Statement II is incorrect, ruling out **(A)** and **(D)**. We are only left with **(B)**, the correct answer. We already know Statement III to be correct, but upon reading, it becomes apparent that this choice presents an entity being adversely affected by forces outside its own borders that it can act upon only if it assumes financial responsibility, or the tying-up of its resources. This situation is analogous to the situation described in paragraph 1.

10. B

This Strengthen–Weaken (Beyond the Passage) question directs us back to paragraph 4 where we must first identify the author's *main contention*. The author argues in this paragraph that *state participation in policy-making must be harmonized with existing federal legislation*. Thus, we need to identify the answer choice that most strongly suggests that state policy-making needn't necessarily be harmonized with federal legislation to weaken this statement. **(A)** does indeed suggest that states can overcome differences with federal legislation; however, only for *minor...discrepancies* and after *lengthy and extremely expensive appeal processes*. Does this weaken the contention that states must be consistent with the federal government? Slightly—it demonstrates that it is possible to achieve victories. Still, it is a very weak refutation of the author because only minor victories can be achieved at a great cost. Further, it doesn't mention whether any of these minor victories were won in pursuit of caring for the national parks. **(B)** offers the point that states can legally appeal the supremacy clause in *any* case pertaining to national parks, implying both major and minor cases. This would have a much more profound effect on the author's argument, weakening it. While this answer choice is the correct one, it is worthwhile to eliminate the others because it would be possible for another answer choice to have a more significant weakening effect. **(C)** actually strengthens the author's contention that states need to be consistent with the federal legislature—otherwise, they'll never achieve anything. Finally, **(D)** has no apparent effect on the author's argument. We don't have enough information to determine if this *revision of [the] mandate* would result in more state power and less need to be consistent with the federal government, so we can eliminate this answer.

Effective Review of CARS

12

12: Effective Review of CARS

In This Chapter

12.1 Learning from Your Mistakes — 304
   Why I Missed It Sheets — 304

12.2 Thinking Like the Testmaker: Post-Phrasing — 306

12.3 Improving Your Timing — 308
   Pacing Guidelines — 308
   Managing Question Timing — 309

12.4 Building Endurance — 310

12.5 Enhancing Your Vocabulary — 310
   Reading Plan — 310

Concept and Strategy Summary — 313

Introduction

> **LEARNING GOALS**
>
> After Chapter 12, you will be able to:
>
> - Troubleshoot common errors on missed questions with Why I Missed It Sheets
> - Apply post-phrasing analysis to difficult passages
> - Manage and adjust your pacing within the CARS section to meet the 90-minute deadline
> - Build endurance and vocabulary through targeted practice

This final chapter is a troubleshooting guide for raising your score. Practice the strategies discussed throughout this book, especially the Kaplan Method for CARS Passages from Chapter 4 of *MCAT CARS Review* and the Kaplan Method for CARS Questions from Chapter 8. Start by working on some practice passages and questions from Full-Length Exams or other materials in your online Kaplan resources. Using these online resources is a prerequisite for any of the review strategies discussed in this chapter to be effective. We'll look at five proven ways to increase your *Critical Analysis and Reasoning Skills* (CARS) section score—including some methods you'll find useful for the science sections as well!

Hands down, the best way to improve is to learn from the mistakes you make on practice tests, which is why we begin our discussion with Kaplan's Why I Missed It Sheets (WIMIS). The post-phrasing strategy discussed in the following section builds on the WIMIS, helping you go the extra mile to begin thinking more like the

MCAT Critical Analysis and Reasoning Skills

writers of the MCAT. Subsequently, we'll talk about a simple approach for managing your pacing on Test Day, and then we'll discuss what you can do to build your test-taking stamina. We'll close with a discussion of one final advantage you can give yourself for CARS: a stronger vocabulary.

12.1 Learning from Your Mistakes

At this juncture, you have read quite a bit about how to approach CARS, have had the opportunity to practice, and may even have completed one or more Full-Length Exams. Now it is time to reflect on your performance and adjust your study plan based on your personal test-taking pathologies, identifiable patterns in your past errors that can help to guide your future CARS studying.

WHY I MISSED IT SHEETS

It can be overwhelming to try to discern what your strengths and areas of opportunity are by looking at an entire CARS section all at once. Rather, take time to analyze each question you answered incorrectly. One way to help manage this process—and to keep a clear record of your performance over time—is to set up **Why I Missed It Sheets** (**WIMIS**). Create a document with four columns: question number, question type, topic (or discipline), and "Why I Missed It." In the first column, make a list compiling all of the items you answered incorrectly on the test. In the second column, identify the question type based on the Kaplan classifications we reviewed in the previous three chapters. The third column should include the topic, or academic discipline, for the passage. Lastly, in column four, take time to review your incorrect response and identify the specific reason why you got the question wrong and missed those points. Table 12.1 demonstrates what WIMIS should look like.

Bridge

Need some help identifying the question type for your Why I Missed It Sheets? Don't forget to review Chapters 9, 10, and 11 of *MCAT CARS Review* for thorough coverage of each question type.

Question Number	Question Type	Topic/Discipline	Why I Missed It
Full-Length X, #12	Detail	Music	Misinterpreted the question stem, leading me to the wrong paragraph for information.
Full-Length X, #21	Strengthen–Weaken (Within)	Art	Chose a piece of evidence that would strengthen the critics' point of view, but was looking to strengthen the author's point of view.
Full-Length X, #38	Inference	Population Health	Chose an answer choice that *may* be true based on the information in the passage but does not have to be—Inference questions have answers that *must* be true based on the passage.
Full-Length X, #53	Main Idea	Psychology	Answer choice was too narrow, only describing two of the paragraphs but not the entire passage.

Table 12.1. Sample Why I Missed It Sheet (WIMIS)

The most important element of this table is the *Why I Missed It* column. Do not just categorize your test-taking mistakes with a generic *I didn't understand* or *careless mistake*. Such comments will not give you insight into why you missed the question. Instead, choose something that is—contradictory as this may sound—as specific and as generalizable as possible. Did you read the question wrong? Did you miss a major point in the passage? Were you unable to see the relevance of a piece of information in the passage or how new information pertained to the passage? Write things that identify the flaw in the thinking process as specifically as possible, while also making sure that you can extrapolate what you learn from that question to future questions. Consider the following pairs of "bad" and "good" examples:

> **Bad:** *I missed this because the answer choice declares that* Beowulf *was written by an aristocrat, but the passage said that the author was only "probably" a member of the royal court.*
> **Good:** *I missed this because the answer choice stated something as fact that the author of the passage did not.*

> **Bad:** *I missed this because I thought the question was asking for the author's opinion on Mendelssohn's work, but it was actually asking for his opinion on the critics of Mendelssohn's work.*
> **Good:** *I missed this because I confused which opinion a question stem was asking for.*

> **Bad:** *I missed this because I recognized "unmoved mover" from the question stem, which is in paragraph 3, but the answer they were looking for actually comes from the description of the "Central Headquarters" sentence in paragraph 4.*
> **Good:** *I missed this because I failed to see how the paragraphs fit together (how an idea in one paragraph is further explained in subsequent paragraphs) before moving on to the questions.*

When reviewing your WIMIS, compare the question type column to the topic/discipline column to narrow the scope of your future study plan. Is there a recognizable pattern you can utilize to optimize your performance? For instance, are you repeatedly missing *Reasoning Within the Text* question types? Or are you only missing this category of question types in Anthropology passages? Use the WIMIS to plan which passage varieties you'll focus on in the future or to determine which chapters you need to reread in *MCAT CARS Review*. You want to make sure that you are moving forward with a plan that specifically supports your areas of opportunity.

If you repeatedly read questions incorrectly, you are not alone—it is extremely common for students to misread or misinterpret questions in CARS, which can lead to selecting an incorrect answer. Making an oversight and missing a word in a question

Real World

In medical school, you'll attend morbidity and mortality (M&M) conferences, in which physicians review clinical and medical errors, patient complications, and other healthcare quality measures to investigate the root causes of problems in order to fix them directly. These conferences are not dissimilar to creating Why I Missed It Sheets—you can only improve if you know what you're doing wrong!

MCAT Critical Analysis and Reasoning Skills

or reading too rapidly and not fully understanding what is being asked can lead to errors. Take your time to read the questions critically to ensure you understand what is being asked. Remember, establishing the question type and finding clues in the question stem about where to research are critical components of the Assess and Plan steps of the Kaplan Method for CARS Questions. As you review your practice tests, reexamine the questions and the answers you selected that were incorrect. In hindsight, when looking at the question a second time, do you find yourself shaking your head because you simply misinterpreted what was being asked? This could indicate that you are reading the question stems too hastily and moving on to the process of answering the question before you even know *what* you're supposed to be answering! If this is the case, force yourself to take the extra time to reword a difficult question stem so that it becomes clear to you precisely what it is asking. This is not wasted time—having more clarity on the question *behind the question* (that is, what the question is *really* asking) will allow you to create a Plan more quickly, Execute that Plan more efficiently, and Answer the question more accurately in less time.

If your WIMIS demonstrates a pattern in which you keep missing the same question types, then start to triage those questions. First, go back and reread the appropriate chapter in *MCAT CARS Review* so you can begin to recognize these question types with greater ease. Chapters 9, 10, and 11 introduced the eight major question types and highlighted the common question stems that fall into each type. Remember that each question type has an associated strategy that will help you accomplish the task or tasks required by the question stem.

Bridge

Knowing your personal test-taking pathologies is essential to improving your score. In addition to looking at the question types, topics or disciplines, and how you read question stems, don't forget to look at Kaplan's classifications of Wrong Answer Pathologies (Faulty Use of Detail, Out of Scope, Opposite, and Distortion), discussed in Chapter 8 of *MCAT CARS Review*.

Finally, take note of the Wrong Answer Pathologies of the incorrect answers that you chose. Simply knowing the Wrong Answer Pathologies that most often sway you from choosing the correct answer is enough to help you avoid falling for the same traps in future tests. For example, if you know that you frequently choose Out of Scope answer choices in Main Idea questions, then ask yourself *Is this answer truly within the confines of the passage, or does it bring in something else?* before finalizing your answer for Main Idea questions in the future.

12.2 Thinking Like the Testmaker: Post-Phrasing

With **post-phrasing**, you go over both why incorrect answers are wrong and why correct answers are right. This strategy is especially helpful if you frequently experience the classic test-taking dilemma: *I can usually narrow it down to two answer choices, and then I always go for the wrong one!*

Post-phrasing begins with your WIMIS. Once you have completed your WIMIS, identify which questions you answered incorrectly and make a note of the correct answer. The post-phrasing process then requires you to articulate in your own words what the question stem says. You may find it helpful to write down your reworded question on a piece of scratch paper or in a word processor. Your focus in this first step should be decoding what the question stem is actually asking you to do. Determine the question type and write or state out loud what the appropriate Plan is for such a question.

Once you've determined the appropriate Plan for the question, Execute the Plan, writing out your prediction for the correct response. It's essential to take the time to write this down: many students find that a nebulous or poorly focused prediction is what keeps them from being able to Answer the question at the end of the Kaplan Method for CARS Questions. The more you practice writing out your predictions during the post-phrasing process, the more routine making predictions will become.

You may notice that the description of post-phrasing, thus far, basically describes a regimented and very deliberate use of the Kaplan Method for CARS Questions. But at the end, the focus will not be on matching your prediction anymore; after all, you already know the correct answer because you're reviewing a question you answered incorrectly. Instead, turn your attention to the incorrect answers. Determine why each incorrect answer is unsuitable. Does it fail to match your prediction? Does it have a common Wrong Answer Pathology? Are there other subtle flaws in logic or reasoning that misrepresent the author's arguments? By taking the time to write out an explanation for each of the incorrect answers, you'll begin to see the correlation between questions and correct answers, as well as reasons for eliminating wrong answers in future questions that you encounter.

Alternatively, post-phrasing can be used as an exercise to improve your recognition of Wrong Answer Pathologies in passages you haven't seen yet. Consider practicing with a few passages following the steps below:

1. Find a fresh CARS passage from your Kaplan resources, as well as its explanations.
2. Go through and select all the correct answers for the passage. Don't read the explanations, just circle and set the answers aside for now.
3. Outline the passage as you normally would.
4. When you work on the questions, do so in a fundamentally different way: as above, the goal here is not to find the right answers (as you have already identified them) but to reason why each incorrect answer is wrong. Read the right answer to be sure you understand it, but spend more time identifying the types of wrong answers presented by the other choices.

MCAT Critical Analysis and Reasoning Skills

> **Bridge**
>
> In addition to Wrong Answer Pathologies, take time to review the Signs of a Healthy Answer presented in Chapter 8 of *MCAT CARS Review*. These include appropriate scope, agreement with the author, and "hedging" language that creates weaker claims.

5. Look at the explanations for each incorrect answer choice and compare them with your notes. If there are any major omissions or other discrepancies in your explanations, be sure to note them prominently. The more you practice, though, the closer you should find the correspondence between your post-phrasing notes and Kaplan's explanations.

It won't take long before you find yourself recognizing the correct answer and thinking *this one just feels right*. Having spent time concentrating on identifying wrong answers, you will begin to see the patterns emerge, and the subtle flaws in these choices make themselves more pronounced. Taking the time to post-phrase will expand your awareness of how the test is written and what the expectations are for CARS questions. Furthermore, you will refine your thought process and your approach to passages and questions, making you better prepared for Test Day.

12.3 Improving Your Timing

Proper management of the clock can make a significant difference to your score in any section, but the timing constraints in the CARS section differ from those of the other sections. Consider a few basic facts about the CARS section:

- It lasts 90 minutes.
- There are 9 passages.
- There are 53 questions.

> **MCAT Expertise**
>
> While it is most common to have five to seven questions associated with a given passage, there have been instances of passages with only four or as many as eight questions. This should not significantly impact how you approach these passages, although—all else being equal—a passage with eight questions will give you more points for the same amount of reading.

It's a simple matter of arithmetic to see that 9 passages in 90 minutes allows you 10 minutes for each passage. Because all of the points actually come from answering the questions, though, you'll want to ensure that you allot them enough time. Passages vary in difficulty, meaning that some will certainly take longer to read than others, but as a rule of thumb you should aim to complete reading and Outlining the passage in about 4 minutes. Subtracting those 36 minutes for reading leaves you with 54 minutes for the 53 questions, which works out to almost exactly one minute per question. Keep in mind that the number of questions associated with a passage can vary from 5 to 7, meaning that a more precise expectation for completing a passage and its questions is somewhere between 9 and 11 minutes.

PACING GUIDELINES

Now, trying to ensure that each passage takes only 4 minutes and each question only one minute would actually be counterproductive: not only would you lose a substantial amount of time checking the clock, you'd likely heighten your anxiety whenever you came across the inevitable question that takes a little extra time to get through. A better approach to managing your time is to check your timing only at a limited number of predetermined points during the section. We recommend checking the clock after every other passage. Assuming that you take 10 minutes to read a passage and answer its questions, you should ideally be no more than 20 minutes further into the section each time you check the clock. We recommend jotting down how much time remains when you start the third, fifth, seventh, and ninth passages (it should be at least 70, 50, 30, and 10 minutes, respectively). If your timing appears to go off-course, note which passages seem to have caused this trouble during your review of the test. Could you have triaged these passages for later and answered more questions correctly in the limited time you had left?

While this approach does not leave you much of a cushion for going back to previous passages and questions, this is by far the easiest way to master pacing. Given that each passage contains between 500 and 600 words, jumping between passages is a far less viable option in CARS than in any of the science sections, where passages can be less than half that length. While it's recommended that you skip passages that you decide in your Scan step will take you too long, we do not recommend giving up in the middle of a passage you decided to work on to return to later. Once you decide to work on a passage, commit to finishing it.

MANAGING QUESTION TIMING

While these guidelines are useful for managing the section, it can still be tricky to master how to split your time between reading the passages and answering the questions. In your online resources, we record the time spent working on each question so you can get a sense of which questions take you longer than others. (Note that this mechanism requires you to click on the question when you begin working on it to accurately record this information.) Keep in mind that the first question of each passage will build in the time you spend critically Reading and Outlining the passage in the beginning. Thus, you should strive to have the first question of each set answered in within 5 minutes (or 300 seconds, to be precise), and every subsequent question should be close to the one-minute mark calculated above.

Once you have a better sense of which types of passages and questions take you longer to complete, you should use this information in conjunction with what you've learned from each WIMIS to guide your *now* or *later* decisions in the first step of each method (Scan for passages and Assess for questions). Doing the passages that are easiest for you first will allow you to get ahead on the timing curve. Similarly, saving the toughest questions for the end of each set will allow you to gain additional familiarity with the passage as you work on its more manageable questions. Because questions can repeat the same theme, you may even find the answer to a challenging question while working on an easier question later in the set.

> **MCAT Expertise**
>
> If the seconds ticking down causes you too much anxiety, an alternative way of working on your timing is to use the timer on your phone or some other timepiece to "time up"—that is, to count upwards from zero to see how much time it's taking you to read the passage and to complete each question. Timers with "lap" functions can be especially useful because these can record how long each question takes. Generally, this method works best with only one or two passages at a time.

12.4 Building Endurance

Preparing for the MCAT is like preparing for a marathon. You cannot run 1 or 2 miles a day for two months and then expect to be successful in a 26.2-mile marathon. Runners vary their running time and build up to the 26.2 miles, routinely running long distances in preparation for marathon day.

How often are you studying and for how long? If you are only studying in 30- or 60-minute increments, it could be that when you sit down to complete a Full-Length Exam that you have not built up your endurance for that type of scenario. Not only do you need to study and practice regularly, but you also need to simulate the Test Day experience of concentrating for 6 hours and 15 minutes of testing time. Gear up for practice tests—and Test Day itself—by periodically holding long study and practice sessions. When working on CARS, try studying in 90-minute increments to mimic the amount of time you'll have for this section.

12.5 Enhancing Your Vocabulary

> **Bridge**
>
> While you will not be expected to know the definition of a piece of jargon from outside knowledge, Definition-in-Context questions hinge on your ability to determine the meaning of a word or phrase from the rest of the passage. These questions are discussed in Chapter 9 of *MCAT CARS Review*.

The AAMC claims that there's no outside content required for the CARS Section of the MCAT, but in truth this is not 100 percent accurate—you may come across a question that has an element of common knowledge (the number of days in a week, for example). Furthermore, while very few questions hinge on knowing the definition of a piece of jargon from outside knowledge, it is nevertheless a tremendous asset to go into Test Day with a strong vocabulary. If nothing else, you can benefit by becoming more comfortable with academic writing in the various disciplines that the AAMC includes in the CARS section by familiarizing yourself with plenty of examples.

The Kaplan and AAMC Full-Length Exams are of course the best place to go to get samples of CARS-style passages and questions. But students who want to go the extra mile also have the option of practicing with outside reading.

READING PLAN

The more often you read, the stronger your reading comprehension skills become, and the faster you evolve into a more efficient reader. This practice will help you build up a wider array of words that you recognize by sight and thus require less time to decipher a text. To prepare for CARS, read on a consistent basis; this will increase your reading speed so that you can maneuver swiftly through passages.

Set up a regimented reading schedule for at least 20 minutes a day, which is just enough time to get through two passages according to the guidelines given above. Or, keep an array of reading materials handy for when you have time available. This mini-library could consist of a list of links kept on a smartphone or other wireless-enabled device. It is surprising how much time we spend waiting—for a friend at a coffee shop, for a group to go out in the evening, for a professor to arrive to class, in a hallway or lounge before a meeting—that could be used productively, sharpening reading skills to prepare for CARS.

When selecting materials, try to simulate the variety of passages you'll encounter on Test Day with a blend of texts from both the humanities and the social sciences. Use your WIMIS to determine which types of passages cause you the greatest confusion or frustration, the ones that slow you down the most on your Full-Length Exams. Focus on reading those types of passages in particular; the more you read these difficult texts, the more familiar you will become with their jargon and other conventions, and the less intimidating these passages will start to seem. Consider using one of the following online services to find academic journal articles in the disciplines that give you the most trouble:

- JSTOR (jstor.org)
- Oxford Journals (academic.oup.com/journals)
- Google Scholar (scholar.google.com)
- Project MUSE (muse.jhu.edu)
- The Directory of Open Access Journals (doaj.org)
- Sage Journals (journals.sagepub.com)

Whenever you come across a word that you don't recognize, stop and take a moment to look that word up, and write down its definition *in your own words*. Keeping a list of these new terms and looking over them periodically will go a long way toward building your vocabulary. As your vocabulary expands, you will be able to recognize words or infer their meaning more swiftly, which will increase your efficiency both with reading and Outlining the passage and with tackling question stems and answer choices.

In addition to learning the meanings of new words, it's also helpful to become acquainted with the major themes and concepts that are distinctive to each of the humanities and social sciences that appear on the CARS section. Familiarity with a concept can allow you to glean more from a passage when it's mentioned, enabling you to read more quickly if the text simply repeats what you've already learned. Be careful, however, not to bring in any ideas that the passage does not include when answering its questions (there's a reason Out of Scope is a common Wrong Answer Pathology)! Because there are only nine passages in CARS, the likelihood that you will have read about the exact same scenario presented in a CARS passage is not high, but it is very likely that you will have read about similar topics.

MCAT Critical Analysis and Reasoning Skills

Additionally, any form of academic reading forces you to think critically about the ideas the author is presenting. Just as when you practice with passages in CARS, try Outlining a few paragraphs of an academic article. Focus on how the author structures the argument: what conclusion does he or she want the audience to reach? What evidence is used to support that conclusion? Are there any flaws in the author's logic? Focused, regular reading will help prepare you for CARS by bolstering your reading comprehension and reasoning skills, as well as your comfort with academic texts and the challenges they bring.

Conclusion

Medical schools want to admit students with strong reasoning skills because higher-order thinking is necessary for both appropriately diagnosing patients and conducting groundbreaking research. When given a constellation of symptoms and concerns, a physician needs to generate a differential diagnosis—a list of the potential ailments described by the symptoms. From this list, the medical team must rule out unlikely diagnoses and provide evidence for the most likely diagnosis. What starts out as a list of a hundred possible causes of headache—from migraines and tension-type headaches to intracranial bleeds and brain tumors—is reduced to one most likely cause after asking appropriate questions (*When did the headache start? Where do you feel the headache? Have you ever had anything similar before?*), performing a thorough physical exam (cranial nerve function, eye exam, looking for evidence of trauma), and running appropriate laboratory and imaging tests (head CT, MRI, inflammatory markers).

Patients want to trust in the expertise of their doctors, and you want to ensure that you are able to deliver the best treatment to your patients. The same skills in CARS that enable you to determine the author's perspective, distinguish her voice from others that appear in the passage, predict a response to a question, and match your predictions (while eliminating incorrect answers) will serve you well as a physician. In the future, you may not be expected to think critically about dance theory, musicology, archaeology, and linguistics, but you will have to synthesize disparate pieces of information, consider assumptions about patient care, and respond appropriately to all parts of your patients' questions—both what they say and what they leave for you to infer. As pointed out in this chapter, the skills tested in the *Critical Analysis and Reasoning Skills* section can always be improved through pointed and actionable review of your past performance. Congratulations on reaching the end of *MCAT Critical Analysis and Reasoning Skills Review*. Though this book now comes to an end, for you this is merely a beginning—good luck on the MCAT, and in all your endeavors in medicine!

CONCEPT AND STRATEGY SUMMARY

Learning from Your Mistakes

- Create **Why I Missed It Sheets** (**WIMIS**) to look for your test-taking pathology patterns.
    - Make a four-column table: question number, question type, topic/discipline, and "Why I Missed It."
    - In the Why I Missed It column, describe the error you made in the question as specifically as possible (with respect to the thought pattern), but in a way that allows you to extrapolate what you learn from that question to future questions.
    - Look for patterns in your WIMIS.
- If you misread questions, be sure to slow down and reword the question stem to make sure you know what question you are actually trying to answer.
- Reread the relevant chapters in *MCAT Critical Analysis and Reasoning Skills Review* as needed.

Thinking Like the Testmaker: Post-Phrasing

- Find a CARS passage from your Kaplan resources, as well as its explanations. This could be a passage you have already read or a new one.
- Go through and circle all the correct answers for the passage. Don't read the explanations—just circle and set the answers aside for now.
- Outline the passage as you normally would.
- When you work on the questions, do so in a fundamentally different way: the goal here is not to find the right answers (as you have already identified them) but to reason why each incorrect answer is wrong. Read the right answer to be sure you understand it, but spend more time identifying the types of wrong answers presented by the other choices.
- Look at the explanations for each incorrect answer choice and compare them with your notes. If there are any major omissions or other discrepancies in your explanations, be sure to note them prominently.

MCAT Critical Analysis and Reasoning Skills

Improving Your Timing

- Aim to read a passage in about 4 minutes; aim to answer each question in about one minute.
- Each passage and its questions together should take somewhere between 9 and 11 minutes.
- Check the clock after every other passage and its questions. Each passage pair should take about 20 minutes.
  - At the beginning of the section, you have 90 minutes left.
  - After two passages, you have about 70 minutes left.
  - After four passages, you have about 50 minutes left.
  - After six passages, you have about 30 minutes left.
  - After eight passages, you have about 10 minutes left.
  - After nine passages, you have 0 minutes left and are finished with the section.

Building Endurance

- Increase stamina by studying in 90-minute increments when possible (equal to the amount of time for the CARS section).
- Periodically, simulate the Test Day experience with study and practice for 6 hours and 15 minutes (equal to the amount of testing time).

Enhancing Your Vocabulary

- Read academic texts for at least 20 minutes a day to sharpen reading skills in preparation for CARS.
- Choose articles on topics that cause you trouble on Full-Length Exams (as revealed by your WIMIS).
- Practice Outlining using these articles.

Notes

Dive Deep into Special Editions

Explore over 50 single-topic special editions from *Scientific American* and *Scientific American MIND*.

From the Science of Dogs & Cats to Physics, our in-depth anthologies focus the lens on a distinct subject in fascinating detail. Previously available on newsstands, these special editions are now reissued in digital format on our website for you to explore.

Find the special edition for you at
scientificamerican.com/collections

    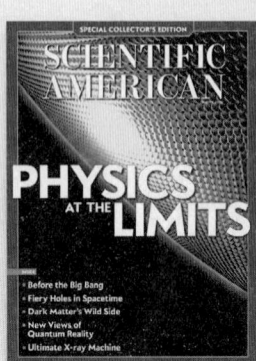

Copyright © 2016 by Scientific American, a division of Nature America, Inc. All rights reserved.